LECTURES ON BUILDINGS

LECTURES ON BUILDINGS

Updated and Revised

Mark Ronan

THE UNIVERSITY OF CHICAGO PRESS • CHICAGO AND LONDON

Mark Ronan is professor emeritus at the University of Illinois at Chicago and honorary professor of mathematics at University College, London. He is the author of *Symmetry and the Monster.*

The University of Chicago Press, Chicago 60637
The University of Chicago Press, Ltd., London
© 1989 by Mark Ronan
All rights reserved. Published 2009
Printed in the United States of America

18 17 16 15 14 13 12 11 10 09   1 2 3 4 5

ISBN-13: 978-0-226-72499-7 (paper)
ISBN-10: 0-226-72499-9 (paper)

Library of Congress Cataloging-in-Publication Data
Ronan, Mark.
Lectures on buildings / Mark Ronan ; Updated and Revised
p. cm.
Includes bibliographical references and index.
ISBN-13: 978-0-226-72499-7 (pbk. : alk. paper)
ISBN-10: 0-226-72499-9 (pbk. : alk. paper)
1. Buildings (Group theory).   2. Finite geometries.
3. Finite groups.   I. Title.
QA179.R64 2009
512′.2—dc22
2009009149

To
Piers and Tamsin

CONTENTS

INTRODUCTION TO THE 2009 EDITION

Since this book was written, twenty years ago, the main developments have been the new theory of twin buildings, which first appeared in print in 1992, and the classification of Moufang polygons by Tits and Weiss, which appeared in a complete form in 2002. A new chapter devoted to twin buildings is now included, along with small amendments to Appendix 2 and the last paragraph of Appendix 1 dealing with Moufang polygons. Up-to-date references for these modifications and for the new Chapter 11 are also included. There have also been small modifications to Chapter 1. It is a pleasure to thank Richard Weiss, Bernhard Mühlherr and Peter Abramenko for their helpful comments on the new chapter and other modifications.

I would also like to thank the various mathematicians who recommended the republication of this book, and the University of Chicago Press for persuading me to go ahead with it. It has been a pleasure to work with them, and in particular I thank my editor Jennifer Howard for her enthusiastic support.

London, July 2008

INTRODUCTION

The genesis of this book was a set of notes taken by students who attended a course of fifteen 2 hour lectures in the University of London at Queen Mary College in 1986/87. After rewriting these notes, I used them in Chicago at the University of Illinois in 1987/88, in a slightly longer course comprising twenty 2 hour lectures. The subsequent expansion and revision of the notes is what appears here, though the appendices largely contain material not covered in the courses. As to the overall structure, the first five chapters deal with the general theory, while Chapters 6-10 cover important special cases. Chapters 2 and 3 are essential to most of what follows, but after that one can be a little more selective. For example a reader wishing to learn about affine buildings and their groups could omit most of Chapters 4, 7 and 8. The Leitfaden which follows gives some idea of the interdependence of the chapters.

A historical account of the origin of buildings is contained in the introduction to the book on spherical buildings by Tits [1974], and I quote, "The origin of the notions of buildings and BN-Pairs lies in an attempt to give a systematic procedure for the geometric interpretation of the semi-simple Lie groups and, in particular, the exceptional groups." Not only has this attempt succeeded, but the theory has been developed far beyond that point, largely by Tits. The term "building", incidentally, is due to Bourbaki.

The buildings for semi-simple Lie groups, and their analogues over arbitrary fields, are of spherical type. Work of Iwahori and Matsumoto [1965] on p-adic groups then led to affine buildings, and the general theory of such buildings, and their groups, has been developed by Bruhat and Tits [1972] and [1984]. Later, Moody and Teo [1972] used Kac-Moody Lie algebras to produce a new class of groups having a BN-Pair, and therefore provided new buildings, of "Kac-Moody type". There is now a class of "Moufang buildings" (Tits [1986], and Chapter 6 section 4) which includes all spherical buildings (having rank ≥ 3 and a connected diagram), and all buildings of "Kac-Moody type"; these include some, but not all, affine buildings (e.g. not the p-adic ones). Moreover these buildings can be constructed independently of the groups (Ronan-Tits [1987], and Chapter 7). There may yet be further interesting classes of buildings, with interesting groups, waiting to be discovered, but certainly the theory has now moved a long way beyond the study of spherical buildings. In fact, affine buildings have been particularly important; they are used for example by Macdonald [1971] in the study of spherical functions on p-adic groups, by Borel-Serre [1976] and Serre [1977/80] in studying arithemetic groups, and by Quillen (see Grayson [1982]) to prove that the K-groups of a curve are finitely generated - see Ronan [1989] for further references.

Finally my thanks are due to all who helped bring this project to fruition: to W.M. Kantor for his excellent lectures on the subject 12 years ago, and his helpful comments on this text; to P. Johnson and S. Yoshiara for very helpful and detailed comments; to J. Tits for some important remarks and suggestions; to Mrs. Ann Cook for typing the first version, and to Ms. Shirley Roper for typesetting the final version. Needless to say the project would never have got underway without the interest of those who attended my lectures in London, and in Chicago: my thanks to all of them and in particular L. Halpenny, M. Iano, M. Mowbray, C. Murgatroyd and M. Whelan who originally took notes in London.

Chicago, September 1988

LEITFADEN

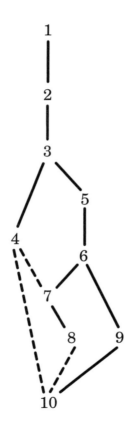

Chapter 1
CHAMBER SYSTEMS AND EXAMPLES

This book treats buildings as chamber systems, and we start by explaining the origin of this concept.

0. Buildings and the Origin of Chamber Systems.

When Tits introduced buildings, he started with the "spherical" ones, treating them as simplicial complexes. A simplicial complex is often thought of as being built from a set of vertices, where some pairs of vertices span edges (simplexes of dimension 1), some triples span triangular faces (simplexes of dimension 2), some quadruples span tetrahedral faces (simplexes of dimension 3), and so on. The faces of each face must be included, so for example the edges of each triangular face are faces of the simplicial complex.

An example of a spherical building—discussed in more detail in Section 3 of this chapter—is the following: take a vector space of dimension $n + 1$, and assign a vertex to each proper subspace. A subset of vertices forms a simplex precisely when the corresponding subspaces form a nested sequence. This implies that a set of vertices forms a simplex if and only if each pair of vertices in that set forms an edge, and Tits [1974] called such a simplicial complex a "flag complex." The simplexes of maximal dimension he called *chambers*, and if every simplex was a face of a chamber, he called this a "chamber complex."

In this context the faces of codimension 1—meaning those of dimension one less than the chambers—Tits called *panels*. In the example above, the chambers are nested sequences of length n: $V_1 \subset V_2 \subset \ldots \subset V_n$, where V_i is a subspace of dimension i. The panels are nested sequences of length $n - 1$, so a subspace of some dimension is missing, and this missing dimension is called the *type* of the panel. There are n different types of panels.

As the theory of buildings developed further, Tits found it more convenient to define buildings in a slightly different way, by starting with the chambers rather than the vertices. He called the resulting object a "chamber system," a concept that appeared in Tits [1981].

When treating a spherical building as a chamber system rather than a simplicial complex, two chambers are said to be *adjacent* if they have a panel in common. Returning to our example above, each panel has one of n possible types, and we say two chambers are *i-adjacent* if they have a panel of type i in common. If a chamber is i-adjacent to two other chambers, then those chambers are i-adjacent to one another, because all three chambers share the same panel. The concept of i-adjacency is therefore an equivalence relation on the set of chambers, and the equivalence classes are in a natural bijective correspondence with the panels of type i.

Given two chambers c and d joined by a sequence $(c = c_0, c_1, \ldots, c_k = d)$ of chambers in which c_{r-1} is i- or j-adjacent to c_r, we say c and d lie in the same *{i,j}-residue*. Being in the same $\{i,j\}$-residue is an equivalence relation, and the equivalence classes are called *{i,j}-residues*. In the example above, these "rank 2" residues, for all possible pairs $\{i,j\}$, are in a natural bijective correspondence with the simplexes of codimension 2, in this case those having $n - 2$ vertices. Allowing three different types of adjacency yields "rank 3 residues", which in the example above correspond to simplexes of codimension 3. The higher the rank of the residue, the higher the codimesion of the simplex, or in other words the lower its dimension.

The point here is that while a simplicial complex is normally regarded as built from the ground up, starting with vertices and moving up to higher dimensional simplexes, when treated as a chamber system it is built from the top down, starting with the chambers and moving down to lower dimensional simplexes.

Let us now formalise the concept of a chamber system and its residues.

1. Chamber Systems.

A set C is a *chamber system* over a set I if each element i of I determines a partition of C, two elements in the same part being called *i-adjacent*. The elements of C are called *chambers*, and if two chambers x and y are i-adjacent we shall often write $x \underset{i}{\sim} y$. In this book I will always be a finite set.

Example 1.   Let G be a group, B a subgroup, and for each $i \in I$ let there be a subgroup P_i with $B < P_i < G$. Take as chambers the left cosets of B, and set

$$gB \underset{i}{\sim} hB \text{ if and only if } gP_i = hP_i.$$

The fundamental nature of this example is exhibited in Exercise 2.

Example 2. In the example above let G be given by generators and relations as $\langle r_i \mid r_i^2 = (r_i r_j)^{m_{ij}} = 1, \forall i, j \in I \rangle$. Set $B = 1$, $Pi = \langle r_i \rangle$. This is a *Coxeter system* and G is called a *Coxeter group*; the next chapter is devoted to the study of such systems.

Further Notation. A *gallery* is a finite sequence of chambers $(c_0, \dots, c_k)$ such that c_{j-1} is adjacent to c_j for each $1 \leq j \leq k$; and we shall always assume $c_{j-1} \neq c_j$. The gallery is said to have *type* $i_1 i_2 \dots i_k$ (a word in the free monoid on I) if c_{j-1} is i_j-adjacent to c_j (there may in general be more than one possible type, though this is not the case for buildings). If each i_j belongs to some given subset J of I, then we call it a *J-gallery*.

We call C *connected* (or *J-connected*) if any two chambers can be joined by a gallery (or *J*-gallery). The *J*-connected components are called *residues of type J*, or simply *J-residues*, and we let *cotype J* mean type $I - J$.

In Example 1, for which chambers are left cosets gB, the *J*-residues correspond to left cosets gP_J where $P_J = \langle P_j \mid j \in J \rangle$.

Notice that every *J*-residue is a connected chamber system over the set J. The *rank* of a chamber system over I is the cardinality of I; the residues of rank 1 are called *panels*, or *i-panels* if of type $\{i\}$, and those of rank 0 (type $\emptyset$) are simply the chambers.

A *morphism* $\phi : C \to D$ between two chamber systems over the same indexing set I will mean a map defined on the chambers and preserving *i*-adjacency for each $i \in I$ (thus if $x, y \in C$ are *i*-adjacent then $\phi(x)$ and $\phi(y)$ are too); the terms *isomorphism* and *automorphism* have the obvious meaning. In Example 1 the group G acts by left multiplication as a group of automorphisms.

Given chamber systems $C_1, \dots, C_k$ over $I_1, \dots, I_k$, their *direct product* $C_1 \times \dots \times C_k$ is a chamber system over the disjoint union $I_1 \cup \dots \cup I_k$. Its *chambers* are all k-tuples $(c_1, \dots, c_k)$ where $c_t \in C_t$, and $(c_1, \dots, c_k)$ is *i-adjacent* to $(d_1, \dots, d_k)$ for $i \in I_t$ if $c_j = d_j$ for $j \neq t$ and $c_t \underset{i}{\sim} d_t$ in C_t.

2. Two Examples of Buildings.

Buildings will be defined in Chapter 3. Here we just give two families of examples, viewed as simplicial complexes. Recall the standard notion of a *simplex*: a 0-simplex is a point, a 1-simplex is a line segment, a 2-simplex is a triangle with interior, etc. More generally an n-simplex is a convex portion of

$\mathbf{R}^n$ spanned by $n + 1$ vertices, and each subset of these vertices spans a *face* of the simplex.

Example 3.   The $A_n(k)$ building Δ.

Let V be an $n + 1$ dimensional vector space over a field k, not necessarily commutative. The *chambers* of Δ are the maximal nested sequences of subspaces.

$$V_1 \subset V_2 \subset \ldots \subset V_n$$

where V_i denotes a subspace of dimension i. Two chambers $V_1 \subset \ldots \subset V_n$ and $V_1' \subset \ldots \subset V_n'$ are *i-adjacent* if $V_j = V_j'$ for all $j \neq i$. This gives a chamber system over $I = \{1, \ldots, n\}$. Notice that a residue of type i corresponds to the set of 1-spaces in a 2-space V_{i+1}/V_{i-1}, or in other words to the points of the projective line over k.

If $J = \{i_1, \ldots, i_r\} \subset I$, the reader should check that a residue of *cotype J* (not *type J*) corresponds to a nested sequence of subspaces (usually called a *flag*)

$$V_{i_1} \subset \ldots \subset V_{i_r} \tag{*}$$

Its chambers are those maximal flags $V_1' \subset \ldots \subset V_n'$ where $V_j' = V_j$ for $j \in J$. In particular the residues of cotype i correspond to the i-dimensional subspaces of V. This building can be realised as a simplicial complex whose vertices are the residues of cotype i for all i in I; these are the vertices of the geometric realisation. The simplexes of dimension $(r - 1)$ are those flags such as (*) above. Figure 1.1 shows the geometric realisation of the $A_2(k)$ building when k is the field of two elements.

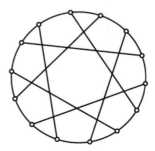

Figure 1.1

A_n **Apartments.**   An important subcomplex of this building, called an *apartment*, is obtained as follows. Fix a basis $v_1, \ldots, v_{n+1}$ of V, and take every sub-

space spanned by a proper subset of this basis, and all nested sequences of
such subspaces. The chambers of the apartment are thus all

$$\langle v_{\sigma(1)} \rangle \subset \langle v_{\sigma(1)}, v_{\sigma(2)} \rangle \subset \ldots \subset \langle v_{\sigma(1)}, \ldots, v_{\sigma(n)} \rangle$$

where σ ranges through all permutations of $1, \ldots, n + 1$. Evidently the symmetric group S_{n+1} acts simple-transitively on the set of $(n + 1)!$ chambers of
this apartment. The reader should note that every panel of this apartment is
a face of exactly two chambers of the apartment. If $n = 2$ and apartment contains six chambers arranged in a circuit; in Figure 1.1 there are 28 apartments. In Figure 1.2 we show an A_3 apartment; it has 24 chambers, 6 on each
face of the tetrahedron. For any n the A_n apartment is the barycentric subdivision of the boundary of an n-simplex (in particular it is a triangulation of
an $(n - 1)$-sphere).

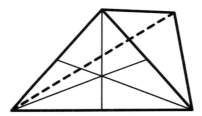

Figure 1.2　　24 chambers - 6 on each face of the tetrahedron.

Example 4.　$C_n(k)$.

　　Let V be a $2n$-dimensional vector space over a commutative field k, endowed with a symplectic form (i.e. a non-degenerate, alternating, bilinear
form). Such a form can be defined on a basis $x_1, \ldots, x_n, y_1, \ldots, y_n$ via:

$$(x_i, y_j) = \delta_{ij} = -(y_j, x_i)$$
$$(x_i, x_j) = 0 = (y_i, y_j)$$

A subspace S is called *totally isotropic* (t.i.) if $(v, w) = 0$ for all $v, w, \in S$; for example $\langle x_1, y_2, y_3 \rangle$. Notice that all 1-spaces are t.i., and that all maximal t.i.
subspaces have dimension n (see Exercise 5). Let $I = \{1, \ldots, n\}$ and for each
$i \in I$ let S_i denote a t.i. subspace of dimension i.

　　A maximal nested sequence

$$S_1 \subset S_2 \subset \ldots \subset S_n$$

of t.i. subspaces is called a *chamber*. As in the previous Example, two cham-
bers $S_1 \subset \ldots \subset S_n$ and $S_1' \subset \ldots \subset S_n'$ are said to be *i-adjacent* if $S_j = S_j'$ for all

$j \neq i$. This is the building $C_n(k)$ as a chamber system. As in Example 3, it can be realised as a simplicial complex by taking the t.i. subspaces as vertices, and taking all t.i. flags as simplexes.

Given the basis $x_i, \ldots, x_n, y_1, \ldots, y_n$ above, we obtain an *apartment* by taking every t.i. subspace spanned by a subset of this basis, and all nested sequences of such subspaces. The chambers of this apartment are thus all

$$\langle v_{\sigma(1)} \rangle \subset \langle v_{\sigma(1)}, v_{\sigma(2)} \rangle \subset \ldots \subset \langle v_{\sigma(1)}, \ldots, v_{\sigma(n)} \rangle$$

where v_j is either x_j or y_j, and σ ranges through all permutations of $1, \ldots, n$. Its automorphism group is the semi-direct product $2^n S_n$ which acts simple-transitively on the set of $2^n n!$ chambers. When we realise the building as a simplicial complex this apartment is a subcomplex isomorphic to the barycentric subdivision of the boundary of a cross-polytope (i.e. the convex polytope whose vertices are precisely the $2n$ unit vectors on the coordinate axes of Euclidean n-space); for $n = 3$ the cross-polytope is the octahedron (Figure 1.3).

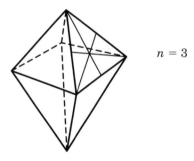

$n = 3$

Figure 1.3   48 chambers - six on each face of the octahedron.

3. Geometric Realisations.

The examples in the previous section were portrayed as simplicial complexes, and all buildings of "spherical type" (see Chapter 6) can be realised in this way. The same is true for buildings whose residues are of spherical type, but when some residues are non-spherical alternative realisations are preferable. A good example is a product of two trees. A single tree having no end points is a building of rank 2. The chambers are the edges of the tree, and are adjacent when they have a vertex in common, the vertices being labelled alternately by two different types. A product of two trees can then be regarded as a chamber system of rank 4, and realised geometrically with the chambers as rectangles, each of which is the product of two edges, one in each tree. The panels are the edges of the rectangles. If we label the vertices of one tree by types 1 and 2, and those of the other tree by types 3 and 4, then each rectangle has edges of all four types, with 1 opposite 2, and 3 opposite 4. The vertices of this realisation represent residues of types $\{1, 3\}$, $\{1, 4\}$, $\{2, 3\}$ and $\{2, 4\}$. The other residues have infinite diameter and are not represented by simplexes.

A different example, also involving tree residues, arises by tiling the hyperbolic plane with right-angled pentagons. This can be treated as a chamber system of rank 5 with the chambers as pentagons, and the panels as edges. Each pentagon has one edge of each type, the edges being indexed by $\mathbf{Z}/5\mathbf{Z}$ in such a way that the four edges meeting at a vertex have labels (i, $i + 1$, i, $i + 1$) in that order. The four chambers meeting at a vertex form a residue of type $\{i, i + 1\}$. The other rank 2 residues can be regarded as trees homeomorphic to the real line, and they and the rank 3 and rank 4 residues all have infinite diameter. This example is a Coxeter system arising from the Coxeter group on five generators r_i, for i in $\mathbf{Z}/5\mathbf{Z}$, and relations given by $(r_i r_{i+1})^2 = 1$ for each i. Coxeter groups are the topic of the next chapter.

Finally if σ denotes a simplex in a geometric realisation of a chamber system, then we let $\mathrm{St}(\sigma)$ denote the corresponding residue, the chambers of $\mathrm{St}(\sigma)$ being the chambers containing σ.

Notes. Chamber systems were introduced by Tits [1981] in "A Local Approach to Buildings", a paper whose main results will be dealt with in Chapter 4. While the spherical and affine buildings in this book can be usefully thought of as simplicial complexes, the examples mentioned above show that it is not always desirable or appropriate to think of a building in that way, and for this reason, and also for the results of Chapter 4, the chamber system formalism is a good way of doing things.

Exercises to Chapter 1

1. Show that the chamber system of Example 1 is connected if and only if $G = \langle P_i \mid i \in I \rangle$.

2. Let C be a chamber system admitting G as a group of automorphisms (i.e. preserving i-adjacency for each $i \in I$) acting transitively on the set of

chambers. Given some chamber $c \in C$, let B denote its stabilizer in G, and let P_i denote the stabilizer of the i-panel on c. Show that C is the chamber system given by cosets of B and the P_i as in Example 1.

3. Let C be the direct product $C_1 \times \ldots \times C_k$ where C_t is a chamber system over I_t. Let x and y be i-adjacent chambers of C, and let X and Y be the I_t-residues containing x and y, where $i \notin I_t$. Show that each chamber of X is i-adjacent to a unique chamber of Y, and i-adjacency gives an isomorphism between X and Y.

4. In Example 3, the group $GL_{n+1}(k)$ acts on V and hence on the building $A_n(k)$; check that this action preserves i-adjacency for each i.

   (i) Show that the stabilizer of a chamber is the subgroup of upper triangular matrices using a suitable ordered basis.

   (ii) Show that any two chambers lie in a common apartment.

   (iii) Show that the subgroup fixing all the chambers of an apartment is the group of diagonal matrices corresponding to a suitable basis.

5. Let V be the $2n$ dimensional vector space of Example 4 having a non-degenerate, alternating, bilinear form. For any subspace W, let $W^\perp = \{v \in V \mid (v, w) = 0 \; \forall w \in W\}$. Show that

$$\dim W + \dim W^\perp = 2n$$

and conclude that all maximal t.i. subspaces have dimension n.

Chapter 2
COXETER COMPLEXES

In this chapter we shall study Coxeter complexes and Coxeter groups. The material here is essential to everything that follows, though only the first three sections will be used in Chapter 3.

1. Coxeter Groups and Complexes.

Let I be a set, and for any $i, j \in I$ let $m_{ij} \in \mathbf{Z} \cup \{\infty\}$ with $m_{ij} = m_{ji} \geq 2$ if $i \neq j$, and $m_{ii} = 1$. The set of such m_{ij} will be denoted by the symbol M. We shall represent M by its *diagram*: the nodes of the diagram are the elements of I (sometimes labelled as such), and between two nodes there is a bond according to the following rule.

$$
\begin{array}{cc}
i & j \\
\circ & \circ \quad \text{no bond if } m_{ij} = 2
\end{array}
$$

$$\circ\!\!-\!\!\!-\!\!\!-\!\!\circ \quad \text{if } m_{ij} = 3$$

$$\circ\!\!=\!\!=\!\!\circ \quad \text{if } m_{ij} = 4$$

$$\circ\!\!\underline{\quad m \quad}\!\!\circ \quad \text{if } m_{ij} = m \geq 5$$

For example the diagram

$$
\begin{array}{ccc}
1 & 2 & 3 \\
\circ\!\!-\!\!\!-\!\!\!-\!\!\circ & \!\!=\!\!=\!\!\circ
\end{array}
$$

means that $m_{12} = 3$, $m_{13} = 2$, $m_{23} = 4$.

The *Coxeter group* of type M is the group W given by generators and relations as:

$$W = \langle r_i \,|\, r_i^2 = (r_i r_j)^{m_{ij}} = 1 \text{ for all } i, j \in I \rangle.$$

For any subset J of I we let W_J denote the subgroup of W generated by all r_j for $j \in J$.

(2.1) LEMMA. *(i) The element $r_i r_j$ in W has order m_{ij}.*

(ii) If $r_i \in W_J$, then $i \in J$.

PROOF: (i) Take a real vector space V having basis $\{e_i | i \in I\}$ indexed by I, and define a symmetric bilinear form on V via

$$(e_i, e_j) = -\cos \frac{\pi}{m_{ij}}$$

In particular $(e_i, e_i) = 1$, and if $m_{ij} = \infty$, then $(e_i, e_j) = -1$. Now for each $i \in I$, let s_i be the linear transformation defined by

$$s_i(v) = v - 2(v, e_i)e_i, \text{ for all } v \in V;$$

and let G be the subgroup of $GL(V)$ generated by the s_i. Let V_{ij} denote the subspace of V spanned by e_i and e_j, and let $V_{ij}^\perp$ denote its orthogonal complement. It is straightforward to check that on V_{ij} the element $s_i s_j$ has order m_{ij} (see Exercise 1), and on $V_{ij}^\perp$ it is the identity. If $m_{ij} = \infty$, this shows $s_i s_j$ has infinite order on V. If $m_{ij} \neq \infty$, then $V = V_{ij} + V_{ij}^\perp$ (Exercise 1), so $s_i s_j$ has order m_{ij} on V. This shows that the map $r_i \to s_i$ extends to a homomorphism of W onto G, and therefore $r_i r_j$ has order m_{ij} in W.

(ii) As j ranges over J, let V_J denote the subspace spanned by the e_j, and let G_J denote the subgroup of G spanned by the s_j. If $r_i \in W_J$, then $s_i \in G_J$, and hence $s_i(v) \in v + V_J$, for all $v \in V$. In particular $-e_i = s_i(e_i) \in e_i + V_J$, so $e_i \in V_J$, and therefore $i \in J$. □

The Coxeter Complex. Take the elements of W as chambers and for each $i \in I$, define i-adjacency by

$$w \underset{i}{\sim} wr_i.$$

This gives a chamber system over I (it is Example 2 in Chapter 1, section 1) and its cell complex is called the *Coxeter complex of type M*; since the r_i generate W it is connected. Notice that each rank 1 residue has exactly two chambers and, by Lemma (2.1), each $\{i, j\}$-residue has $2m_{ij}$ chambers because r_i and r_j generate a dihedral group of that order. The cell complex of a rank 2 residue is thus a polygonal graph; one sometimes thinks of an $\{i, j\}$-residue as being the set of incident point-line pairs of an m_{ij}-gon, two such being i-adjacent (or j-adjacent) if they share a common point (or line) - indeed the dihedral group D_{2m} acts simple-transitively on the set of incident point-line pairs of a regular m-gon.

Examples. For the diagrams A_3(∘———∘———∘) and
C_3(∘———∘════∘) the cell complex is a triangulation of the 2-sphere,
illustrated in Chapter 1 (Figures 1.4 and 1.5). Here are two further examples: the chambers are triangles, and the three types of adjacency are
illustrated by the different types of edges.

Diagram $\widetilde{A}_2$.

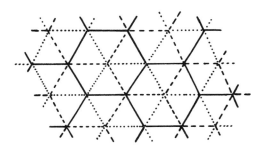

Figure 2.1

Diagram $\widetilde{C}_2$. ∘════∘════∘

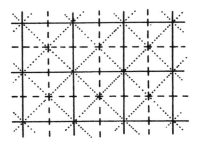

Figure 2.2

    Throughout these notes we shall use W to denote both the Coxeter
group, and the Coxeter complex. As in Chapter 1, an *automorphism* of
a chamber system is a bijective map on the set of chambers preserving i-adjacency for each i. A group action on a set X is called *simple-transitive*
if it is transitive and the stabilizer of $x \in X$ is the identity.

(2.2) LEMMA. *The automorphism group of the Coxeter Complex is the Coxeter group, and it acts simple-transitively on the set of chambers.*

PROOF: Clearly the action of W on itself by left multiplication preserves i-adjacency. On the other hand if we fix one chamber we fix all chambers adjacent to it because each rank 1 residue has exactly two chambers. By connectivity we therefore fix all chambers, and simple-transitivity follows. □

2. Words and Galleries.

Given a word $f = i_1 \ldots i_k$ in the free monoid on I, we set $r_f = r_{i_1} \ldots r_{i_k} \in W$; if $\emptyset$ denotes the null word, $r_\emptyset = 1$. Given $x, y \in W$, notice that there is a gallery of type f from x to y if and only if y can be written as $x r_f$ (the gallery being $(x, x r_{i_1}, x r_{i_1} r_{i_2}, \ldots)$), or equivalently $x^{-1} y = r_f$. For distinct $i, j \in I$ with m_{ij} finite, we write $p(i, j)$ to mean $\ldots ijij$ (m_{ij} factors); e.g. if $m_{ij} = 3$ then $p(i, j) = jij$.

An *elementary homotopy* is an alteration from a word of the form $f_1 p(i, j) f_2$ to the word $f_1 p(j, i) f_2$. Two words are called *homotopic* if one can be transformed into the other by a sequence of elementary homotopies, and we write $f \simeq g$ to mean f and g are homotopic. Notice that two homotopic words necessarily have the same length.

An *elementary contraction* (or *expansion*) is an alteration from a word of the form $f_1 ii f_2$ to the word $f_1 f_2$ (or from $f_1 f_2$ to $f_1 ii f_2$).

We now define two words to be *equivalent* if one can be transformed into the other by a sequence of elementary homotopies, expansions and contractions.

(2.3) LEMMA. *Two words f and g are equivalent if and only if $r_f = r_g$.*

PROOF: Since $r_i^2 = 1$, the relation $(r_i r_j)^{m_{ij}} = 1$ is equivalent to the relation $r_{p(i,j)} = r_{p(j,i)}$, and so the result is immediate from the presentation of W in terms of generators and relations. □

A word is called *reduced* if it is not homotopic to a word of the form $f_1 ii f_2$. Notice that each equivalence class contains a reduced word. We will show later (2.11), that *if two reduced words are equivalent then they are homotopic.*

Example. Consider the diagram $\tilde{A}_2$ (i.e., $m_{12} = m_{13} = m_{23} = 3$), as in the Examples above. Using the theorem (2.11) just alluded to, it

follows that the Coxeter group is infinite, because a word of the form
1 2 3 1 2 3 1 2 3 ... is reduced, and such a word may be arbitrarily long.

A gallery $(x = x_0, x_1, \ldots, x_k = y)$ is said to have *length* k, and the
distance $d(x, y)$ between x and y is the least such k; a gallery from x to y is
called *minimal* if its length is $d(x, y)$. Given $w \in W$ we define the *length* of
w to be $\ell(w) = d(1, w)$, the length of a minimal gallery from 1 to w; notice
that $d(x, y) = d(1, x^{-1}y) = \ell(x^{-1}y)$.

(2.4) LEMMA. *If y' is adjacent to, and distinct from, y, then $d(x, y') = d(x, y) \pm 1$.*

PROOF: If f and g are the types of two galleries from x to some chamber
z, then $r_f = x^{-1}z = r_g$, so by (2.3) f and g are equivalent, and hence
both galleries have even length or both have odd length. Since a gallery of
length k from x to y extends to one of length $k + 1$ from x to y', we see
that $d(x, y)$ and $d(x, y')$ cannot both be even or both be odd. Therefore
$d(x, y) \neq d(x, y')$, and the result follows. □

Reflections and Walls. A *reflection* r is by definition a conjugate of
some r_i; its *wall* M_r consists of all simplexes (of the Coxeter complex)
fixed by r (acting on the left of course). A panel lies in M_r if and only if its
two chambers are interchanged by r, and since the reflection $r = wr_iw^{-1}$
interchanges the i-adjacent chambers w and wr_i, M_r is a subcomplex of
codimension 1.

Notice that if π is any i-panel and x is one of the two chambers on
π, then xr_i is the other chamber on π, and $r = xr_ix^{-1}$ is the unique
reflection interchanging x and xr_i. Thus each panel lies on a unique wall,
and there is a bijective correspondence between the set of walls and the set
of reflections.

We shall say that a gallery $(c_0, \ldots, c_k)$ *crosses* M_r whenever r interchanges c_{i-1} with c_i, for some i, $1 \le i \le k$. We will show that M_r splits
W into two parts interchanged by r.

(2.5) LEMMA. *(i) A minimal gallery cannot cross a given wall twice.*

*(ii) Given chambers x and y, the number of times mod 2 that a gallery
from x to y crosses a given wall is independent of the gallery (i.e., it
is either even for each gallery, or odd for each gallery).*

PROOF: (i) If a minimal gallery $\gamma = (c_0, \ldots, c_k)$ crosses M_r twice, at $(i - 1, i)$ and $(j-1, j)$, then the reflection r sends the subgallery $(c_i, \ldots, c_{j-1})$ to a gallery of the same length from c_{i-1} to c_j. This contradicts the minimality of γ.

(ii) Given $r_f = x^{-1}y$, let $n(f)$ be the number of times the gallery of type f from x to y crosses the wall M_r. If $r_f = r_g$, then by (2.3) f and g are equivalent. If they are equivalent via an elementary homotopy then $n(f) = n(g)$. Indeed an elementary homotopy takes place in a rank 2 residue R, so if the wall M_r contains a panel of R then it actually meets R in two opposite panels (because a reflection fixes two opposite panels in a polygon), in which case both galleries cross M_r exactly once in R. If g is equivalent to f via an elementary expansion or contraction then $n(g) = n(f)$ or $n(f) \pm 2$. $\square$

Let us temporarily call a gallery *even* or *odd* depending on whether it crosses the wall M_r an even or odd number of times. The preceding Lemma (2.5) implies that a given chamber c partitions W into two parts according to the parity of a gallery from c. Given another chamber c', the same partition is achieved, as the reader may readily verify, although there is a switch of parity if a gallery from c to c' is odd. These two parts of W are called the *roots* (or *half-apartments*) determined by the wall M_r. They form complementary subsets of W, and are said to be *opposite* one another; if one is denoted α, the other is denoted $-\alpha$, and if r is the reflection we let $\pm \alpha_r$ denote the two roots.

Before stating the next proposition we define a set X of chambers to be *convex* if any minimal gallery between two chambers of X lies entirely in X.

(2.6) PROPOSITION. *(i) Roots are convex.*

(ii) If α is a root, and x, y adjacent chambers with $x \in \alpha$ and $y \in -\alpha$, then

$$\alpha = \{c | d(x,c) < d(y,c)\}.$$

(iii) There are bijective correspondences between the set of reflections, the set of walls, and the set of pairs of opposite roots.

PROOF: (i) If $c, c' \in \alpha_r$ then by (2.5) a minimal gallery from c to c' does not cross M_r. Thus every chamber on this gallery lies in α_r.

(ii) If $c \in \alpha = \alpha_r$, then by (2.5) a minimal gallery from x to c cannot cross M_r, and hence cannot go via y, so by (2.4) $d(x,c) < d(y,c)$. Conversely, if $d(x,c) < d(y,c)$, then since x and y are adjacent there is a minimal gallery from y to c via x, and this crosses M_r, so $c \notin -\alpha$.

(iii) A reflection determines a wall, and since a given panel is fixed by only one reflection, the wall determines the reflection. Moreover a wall M determines two opposite roots $\pm\alpha$ as above, and if $x \in \alpha$ and $y \in -\alpha$ share a panel π, then by (2.5) and the definition of $\pm\alpha$, the minimal gallery (x, y) crosses M, so $\pi \in M$. Since π determines M, this shows that two opposite roots are associated to a unique wall. $\qquad\square$

Foldings. Let α be any root, and r the corresponding reflection; using (2.6)(ii) one sees that r switches α and $-\alpha$. Thus one has a map

$$\rho_\alpha : W \to \alpha$$

defined by $\rho_\alpha(x) = x$ if $x \in \alpha$, and $\rho_\alpha(x) = r(x)$ if $x \notin \alpha$. It is a morphism (i.e. preserves i-adjacency for each i), because if $x \in \alpha$ is adjacent to $y \notin \alpha$, then clearly $\rho_\alpha(y) = \rho_\alpha(x) = x$. This ρ_α is called the *folding* of W onto α.

The wall M_r determined by α will be denoted $\partial\alpha$ because it is the boundary of α in the usual sense (see Exercise 6). Since any gallery γ from a chamber $c \in \alpha$ to $d \in -\alpha$ crosses the wall $\partial\alpha$, its image $\rho_\alpha(\gamma)$ contains at least one repeated chamber, and hence there is a shorter gallery from c to $\rho_\alpha(d)$; this fact will be used later.

(2.7) PROPOSITION. *Let x and y be chambers, and $(x = x_0, x_1, \ldots, x_k = y)$ any minimal gallery from x to y. For $i = 1, \ldots, k$ let β_i denote the root containing x_{i-1} but not x_i; these β_i are mutually distinct and are precisely all roots containing x but not y. In particular $d(x, y)$ equals the number of roots containing x but not y.*

PROOF: If a root β contains x but not y, then any minimal gallery from x to y goes from β to $-\beta$ at some point, and hence β is one of the β_i. By convexity (2.6) a minimal gallery cannot enter and exit from a given root, so $x \in \beta_i$, $y \notin \beta_i$ and the β_i are distinct. $\qquad\square$

Example. Figure 2.3 shows three minimal galleries from x to y in the $\widetilde{A}_2$ Coxeter complex. Each of these galleries determines an ordering of the roots containing x but not y; these are:

$\beta_1\beta_2\beta_3\beta_4\beta_5$, $\beta_1\beta_2\beta_5\beta_4\beta_3$ and $\beta_5\beta_2\beta_1\beta_4\beta_3$.

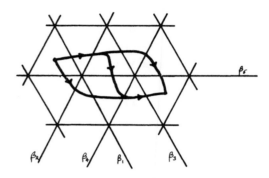

Figure 2.3

(2.8) PROPOSITION. *Given chambers x and y, a chamber lies on a minimal gallery from x to y if and only if it lies in every root containing x and y.*

PROOF: By convexity (2.6) any chamber lying on a minimal gallery from x to y lies in every root containing x and y. Conversely suppose z is contained in every such root. If α is a root containing x but not z, then by hypothesis y is not in α; and if β is a root containing z but not y then again by hypothesis β contains x. Any root containing x but not y is one of the α or β, hence by (2.7) $d(x,z) + d(z,y) = d(x,y)$, so z lies on a minimal gallery from x to y. □

Remark. If in the preceding proposition there are no roots containing both x and y, then every chamber lies on a minimal gallery from x to y. In this case (2.7) implies that W has only finitely many roots, and its diameter is finite. This implies (Exercise 5) that W is finite.

(2.9) THEOREM. *Given any $w \in W$ and any residue R, there is a unique chamber of R nearest w (call it $\operatorname{proj}_R w$), and for any chamber $x \in R$, there is a minimal gallery from w to x via $\operatorname{proj}_R w$.*

PROOF: If b,c are distinct chambers of R at minimal distance from w, take a root containing one but not the other. Without loss of generality this gives a root α with $w, c \in \alpha$, $b \notin \alpha$. If γ is a minimal gallery from w to b, then it crosses from α to $-\alpha$, and hence $\rho_\alpha(\gamma)$ gives a shorter gallery from w to $\rho_\alpha(b) = b'$. However $\rho_\alpha(c) = c$ implies $\rho_\alpha(R) \subset R$, so $b' \in R$. This contradicts the minimality of $d(w,b)$, proving that $\operatorname{proj}_R w$ exists.

To prove the last statement of the theorem it suffices, by (2.8), to show that if α is any root containing w and x, then α contains $\text{proj}_R w$. Since $x \in \alpha$ one has $\rho_\alpha(R) \subset R$, and if $\text{proj}_R w \notin \alpha$, then $\rho_\alpha(\text{proj}_R w) \in R$ is nearer w than $\text{proj}_R w$ is, a contradiction. $\square$

(2.10) LEMMA. *If x and y are chambers in a common J-residue, then any minimal gallery from x to y is a J-gallery. In particular, residues are convex.*

PROOF: Let R be the J-residue concerned, and suppose z lies on a minimal gallery from x to y. If $z \notin R$, set $z' = \text{proj}_R z$; by (2.9) $d(x, z') < d(x, z)$ and $d(z', y) < d(z, y)$, contradicting the minimality of a gallery from x to y via z. Thus $z \in R$. Hence any minimal gallery from x to y lies in R, and it remains to show that if $x, x' \in R$ are i-adjacent, then $i \in J$, but this follows from (2.1)(ii). $\square$

3. Reduced Words and Homotopy.

We observed earlier, following Lemma (2.1), that the $\{i, j\}$-residues of a Coxeter complex have $2m_{ij}$ chambers arranged in a circuit. If x and y are chambers in such an $\{i, j\}$-residue joined by a gallery of type $p(i, j)$, then they are also joined by a gallery of type $p(j, i)$. Thus an elementary homotopy of words $f = f_1 p(i, j) f_2 \simeq f_1 p(j, i) f_2 = f'$ can be realized at the gallery level by making an alteration in some $\{i, j\}$-residue. Recall that a word f is reduced if it is not homotopic to a word of the form $f_1 ii f_2$. As promised earlier we now prove:

(2.11) THEOREM. *A gallery of type f is minimal if and only if f is reduced. Moreover any two reduced words f and g which are equivalent (i.e. $r_f = r_g$) must be homotopic.*

PROOF: The proof consists of two main steps.

Step 1. If f_1 and f_2 are the types of two minimal galleries from x to y, then $f_1 \simeq f_2$.

Let f_1 end in i, and f_2 in j. If $i = j$, then $f_1 = f_1' i$ and $f_2 = f_2' i$, so f_1' and f_2' are the types of two minimal galleries with the same extremities. Induction on the length of the gallery shows $f_1' \simeq f_2'$, hence $f_1 \simeq f_2$. Now suppose $i \neq j$; let R be the $\{i, j\}$-residue containing y, let $z = \text{proj}_R x$, and let y_1 and y_2 be the chambers respectively i- and j-adjacent to y. By (2.9) there are minimal galleries from x via z to y_1 and y_2 respectively, and these extend (by one chamber) to galleries from x via z to y - see Figure 2.4.

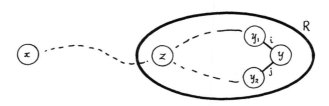

Figure 2.4

By (2.10) the subgalleries from z to y are $\{i,j\}$-galleries, and since R is a $2m_{ij}$-gon, by (2.1), these sub-galleries have types $p(j,i)$ and $p(i,j)$ respectively. Thus if f_0 is the type of some minimal gallery from x to z, then there exist galleries of types $f_0p(j,i)$ and $f_0p(i,j)$ from x to y. By induction, as above, $f_1 \simeq f_0p(j,i) \simeq f_0p(i,j) \simeq f_2$.

Step 2. If f is a reduced word then any gallery of type f is minimal.

Again by induction we assume this to be true if the length of f is less than k (for $k = 0$ the result is trivial). Now let $f = gij$ $(i,j \in I)$ be reduced, and $\gamma = (x_0, \ldots, x_k)$ a gallery of type f. By induction $\gamma_1 = (x_0, \ldots, x_{k-1})$ is minimal. If γ is not minimal, then $d(x_0, x_k) = k - 2$, so there exists a minimal gallery γ_2 from x_0 to x_{k-1} via x_k - see Figure 2.5.

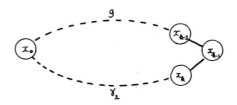

Figure 2.5

Since γ_1 has type gi and γ_2 has type hj, for some word h, we apply Step 1 to see that $f = gij \simeq hjj$ is not reduced. This contradiction shows γ is minimal, as required.

To conclude the proof of the theorem, notice that a minimal gallery must have reduced type otherwise we could replace it by a gallery in which a repeated chamber occurs; the converse is given in Step 2. Now any two reduced words f and g which are equivalent give minimal galleries from 1 to $w = r_f = r_g$, and hence by Step 1, f and g are homotopic. □

(2.12) COROLLARY. *If f_1 and f_2 are reduced words and $f_1 f \simeq f_2 f$ (or $f f_1 \simeq f f_2$), then $f_1 \simeq f_2$.*

PROOF: Indeed $r_{f_1} r_f = r_{f_1 f} = r_{f_2 f} = r_{f_2} r_f$, so $r_{f_1} = r_{f_2}$ and the result is immediate from (2.11). □

(2.13) COROLLARY. *If f is reduced and fj (or jf) is not reduced, then f is homotopic to some word ending (or beginning) with j.*

PROOF: Let g be a reduced word such that $r_g = r_{fj}$. If f has length k, then g has length $k - 1$ by (2.4); and if gj is not reduced then $r_f = r_{gj}$ has length $k - 2$, a contradiction. Therefore gj is reduced and $f \simeq gj$ by (2.11). The jf case follows by symmetry. □

If J is a subset of I we let M_J denote (m_{ij}) for $i, j \in J$.

(2.14) COROLLARY. *The subgroup $W_J = \langle r_j | j \in J \rangle$ of W is the Coxeter group of type M_J.*

PROOF: It suffices to show that an equivalence between two words f and g (i.e., $r_f = r_g$) in the free monoid on J can be realized using only elements of J (i.e., W_J inherits no further relations from W).

From our definition of a reduced word, f and g can be turned into reduced words f' and g' by means only of elementary homotopies and contractions (i.e. without using any elementary expansions), and therefore without using elements outside J. Moreover by (2.1) f' and g' are homotopic, so f and g are equivalent via a sequence of elementary equivalences involving only elements in J. □

4. Finite Coxeter Complexes.

If W is a finite Coxeter complex, let diam(W) denote its *diameter*, the maximum distance between two chambers, and define two chambers to be *opposite* if the distance between them is diam(W). Notice that W necessarily has finite rank (cf. Exercise 5).

(2.15) THEOREM. *If W is finite, then:*

(i)   diam(W) = $\frac{1}{2}$ (no. of roots of W).

(ii) Two chambers are opposite if and only if they lie in no common root.

(iii) Every chamber has a unique opposite.

(iv) If x and y are opposite chambers, then every chamber lies on a minimal gallery from x to y.

PROOF: We first claim that if x and y lie in a common root then they cannot be opposite. Indeed if α is a root containing x and y, set $y' = \rho_{-\alpha}(y)$. Then $d(x, y') > d(x, y)$ because a minimal gallery γ from x to y' must cross the wall $\partial \alpha$, so $\rho_\alpha(\gamma)$ contains a repeated chamber and hence gives a shorter gallery from x to y. Thus x and y are not opposite.

To prove (i) notice first that $\text{diam}(W) \leq \frac{1}{2}$ (no. of roots of W) by (2.7). On the other hand if x and y are opposite in W, then by the above, no root containing x can contain y, and therefore $\text{diam}(W) = d(x, y) \geq \frac{1}{2}$ (no. of roots of W), again by (2.7). This proves (i).

To prove (ii) it remains to show that if x and y lie in no common root then they are opposite; but in this case (2.7) implies $d(x, y) \geq \frac{1}{2}$ (no. of roots of W) so the result follows from (i).

To prove (iii), suppose y and z are distinct chambers opposite x. Take a root α containing one but not the other; either α or $-\alpha$ contains x, so without loss of generality x and y lie in a common root, contradicting (ii). By definition at least one chamber has an opposite in W and hence by transitivity of the group they all do.

(iv) We have shown x and y lie in no common root, so this is immediate from (2.8).   □

Sphericity. A Coxeter complex which is finite is often called *spherical*, or *of spherical type* because the geometric realisation of a finite Coxeter complex of rank n is a triangulation of the $(n-1)$-sphere. The most useful way of seeing this is to use the real vector space V defined in the proof of (2.1); for more details of the following facts see Tits [1968] and Bourbaki [1968/81]. The Coxeter group W acts faithfully on V, and the fixed points for each reflection of W form a hyperplane of V. When W is finite it obviously acts discretely on V, and these hyperplanes partition V into open sets called *Weyl chambers*, each of which is a fundamental domain for W. The reflection hyperplanes intersect a sphere S^{n-1} centreed at the origin of V to give a triangulation of V, which may be identified with the Coxeter complex. Each reflection hyperplane H meets S^{n-1} in a wall M of this Coxeter complex, and the two half-spaces on either side of H correspond to the roots having boundary M. Finiteness of W corresponds to the case of the symmetric bilinear form $(e_i, e_j) = -2\cos(\pi/m_{ij})$ being positive definite, so finite Coxeter groups can be classified by considering these forms. An alternative mode of classification is given in Exercises 9-12.

Observation. Writing $\ell(wr_i) < \ell(w)$ is another way of stating that there is a minimal gallery, whose type ends in i, from 1 to w.

(2.16) THEOREM. *Suppose $\ell(wr_j) < \ell(w)$ for all $j \in J$. Let R be the J-residue containing w, and let $z = \operatorname{proj}_R 1$ be the unique chamber of R nearest 1. Then R is finite and z is opposite w in R.*

PROOF: It suffices to show that every chamber of R lies on a minimal gallery from z to w. Indeed in this case (2.8) implies that z and w lie in no common root of R, hence by (2.7) R has finitely many roots and hence finite diameter; therefore R itself is finite by Exercise 5, and by (2.15) (ii) z and w are opposite. Consider first the case $|J| = 2$. In this case the two chambers of R adjacent to w are closer to 1, and hence closer to z, than w is. Therefore z and w are opposite in the $2m_{ij}$-gon R, and every chamber of R lies on a minimal gallery from z to w (a fact we shall use below).

For the general case, assume that every chamber of R at distance $< k$ from w lies on a minimal gallery from z to w; this is true by hypothesis if $k = 2$. Now let $v \in R$ be at distance $k \geq 2$ from w in R on a minimal gallery $(w, \ldots, v'', v', v)$ of type $\ldots ij$, where $i, j \in J$ - see Figure 2.6).

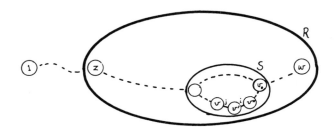

Figure 2.6

Let S be the $\{i,j\}$-residue containing v, and $v_0 = \text{proj}_S w$, so $d(w, v_0) \leq d(w, v'') = k - 2$. The two chambers which are i- and j-adjacent to v_0 are at distance at most $k - 1$ from w, hence by induction lie on minimal galleries from z to w; in particular they are both closer to z than v_0 is. We may therefore apply the case $|J| = 2$, in which S takes the place of R, v_0 takes the place of w, and z takes the place of 1. Thus v_0 is opposite $\text{proj}_S z$ in S, and so v lies on a minimal gallery from $\text{proj}_S z$ to $v_0 = \text{proj}_S w$, hence from z to w as required.   □

5. Self-Homotopy.

The purpose of this last section of Chapter 2 is to prove a theorem which will be applied in Chapters 4 and 7; the details could be omitted at a first reading. For notational convenience we now let $\simeq$ mean only *elementary homotopy*.

A *self-homotopy* is a sequence of elementary homotopies beginning and ending with the same word. Given a word f we let $H(f)$ denote the graph whose vertices are words homotopic to f, and whose edges are elementary homotopies. A self-homotopy is then a circuit in this graph.

Let us call a self-homotopy *inessential* if it is of the form

$$f = f_0 \simeq f_1 \simeq \ldots \simeq f_{k-1} \simeq f_k \simeq f_{k-1} \ldots \simeq f_1 \simeq f_0 = f$$

i.e., "do then undo" - a degenerate circuit in $H(f)$; or if it is of the form

$$f_1 p(i, j) f_2 p(k, l) f_3 \simeq f_1 p(j, i) f_2 p(k, l) f_3$$
$$\simeq \qquad\qquad\qquad \simeq$$
$$f_1 p(i, j) f_2 p(l, k) f_3 \simeq f_1 p(j, i) f_2 p(l, k) f_3$$

"do then undo in reverse order".

We shall say that a circuit π in a graph *decomposes* into two circuits $\pi_1 \pi_2$ and $\pi_2^{-1} \pi_3$ if $\pi = \pi_1 \pi_3$ (here π^{-1} means π in reverse order). This definition extends to the decomposition of a circuit into finitely many circuits, or a self-homotopy into finitely many self-homotopies.

(2.17) THEOREM. *Every self-homotopy decomposes into self-homotopies each of which is inessential or lies in a rank 3 residue of spherical type (i.e. type J with W_J finite).*

PROOF: By induction on the length of the word f, we may assume it is true for words of shorter length than f.

We shall show that a sequence of elementary homotopies of the form

$$fi \simeq \ldots j \simeq \ldots j \simeq \ldots \quad \ldots \simeq \ldots j \simeq gk \qquad (*)$$

can be replaced by one of the form

$$fi \simeq \ldots i \simeq \ldots \quad \ldots \simeq \ldots i \simeq \ldots k \simeq \ldots \quad \ldots k \simeq gk \qquad (**)$$

by decomposing into circuits which are either inessential or else lie in the $\{i, j, k\}$-residue R containing $w = r_{fi} = r_{gk}$.

Note first that by (2.16) R is of spherical type, and $w = w_1 w_2$ where $w = \text{proj}_R 1$ is the unique element of R of shortest length, and w_2 is the longest element of $W_{\{i,j,k\}}$. Let us write $w_1 = r_h$ for some reduced word h, and $w_2 = r_{h'}$ where h' may be chosen to end in i, j, or k. Applying (2.16) to the $\{i, j\}$-residue S of R containing w, we see that h' is homotopic to $h_k p(i, j)$, where h_k is some reduced word such that $r_{h_k} = \text{proj}_S 1$. Similarly h' is homotopic to $h_i p(j, k)$ and $h_j p(k, i)$, with h_i and h_j suitably defined. By (2.12) a homotopy between $\ldots l$ and $\ldots l$ can be done using only words ending in l, so we may alter the original sequence (*) as follows:

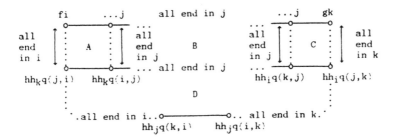

Now, circuits A and C decompose into inessential circuits; B decomposes as required because all terms end in j and we may apply the induction hypothesis; and D involves only self-homotopies in the rank 3 residue R of spherical type. Note that if $i = k$, then $hh_k p(j, i) = hh_i p(j, k)$ and the bottom path reduces to a point.

Finally by using alterations as above from (*) to (**) we may decompose any circuit to one all of whose terms end in i, and then the result follows by induction. □

Notes. Coxeter groups were first studied in complete generality by Tits [1968], and many of the results on Coxeter complexes in this chapter are taken from Tits [1974], though the material in section 5 is from Tits [1981]. Those which act discretely on Euclidean space, namely the finite ones (of spherical type) and those of affine type (Chapter 9) were classified by Coxeter [1934]; see also the elegant paper by Witt [1941]. This classification also appears in the book on Regular Polytopes by Coxeter [1947] which contains a wealth of historical detail and an extensive bibliography; for example, Coxeter remarks that polyhedra of types E_6, E_7 and E_8 were constructed in 1897 by Thorold Gosset, a lawyer practising in London - see [loc. cit.] pp. 202 and 164. All finite Coxeter groups satisfying the crystallographic condition (i.e. all $m_{ij} = 2, 3, 4$ or 6) appear as Weyl groups of semisimple Lie algebras. For more details on this, see Bourbaki [1968/81], particularly the historical sketch on pages 234-240; this book also contains an excellent account of Coxeter groups in the general case.

Exercises to Chapter 2

1. Using the notation of Lemma (2.1) show that $s_i s_j$ has order m_{ij} on V_{ij} and is the identity on $V_{ij}^{\perp}$. If m_{ij} is finite show that $V_{ij} + V_{ij}^{\perp} = V$ and show that this can fail for $m_{ij} = \infty$. [HINT: if $m_{ij} < \infty$ identify V_{ij} with $\mathbf{R}^2$ (having the usual dot product) in such a way that e_i and e_j are unit vectors and $\pi - \pi/m_{ij}$ is the angle between them].

2. Show that a gallery is minimal if and only if it crosses no wall twice.

3. Show that W_J (see (2.14)) is the stabilizer of the J-residue (of the Coxeter complex) containing 1, and that $W_J \cap W_K = W_{J\cap K}$, and $\langle W_J, W_K \rangle = W_{J\cup K}$. In particular, when I is finite, the geometric realisation of W is a simplicial complex, because a simplex wW_J is uniquely determined by its vertices, namely the wW_K, where $J \subset K$ and $K = I - \{i\}$ for some i.

4. (*The Exchange Property*) Let $f = i_1 \ldots i_n$ and suppose $\ell(r_f) > \ell(r_{fi})$. Prove that $r_{fi} = r_g$ where $g = i_1 \ldots \hat{i}_j \ldots i_n$ (i_j removed for some j). [HINT: Let α be the root containing r_{fi} but not r_f, and let γ be the gallery of type f from 1 to r_f; consider the gallery obtained by applying the folding ρ_α to γ].

5. If the diameter of W is finite show that I is finite, and then show W is finite.

6. Let α be a root, and r the reflection switching α with its opposite $-\alpha$. Treating α as a subcomplex of the geometric realisation, by including all faces of chambers in α, show that its boundary $\partial\alpha$ is the wall M_r fixed by r, and that $\partial\alpha = \alpha \cap (-\alpha)$.

7. Let $M_1, \ldots, M_k$ be the connected components of the diagram M, and let I_t denote the nodes of M_t (so I is the disjoint union $I_1 \cup \ldots \cup I_k$). Writing $W_t = W_{I_t}$, show that W is isomorphic to $W_1 \times \ldots \times W_k$ both as a group and as a chamber system. (The W_t are called the *irreducible components* of W.)

8. Give all possible reduced words f such that r_f is the longest word for the A_3 diagram o———o———o (there are 16 of them), and exhibit an inessential self-homotopy (cf. Chapter 8, section 1).

9. If W is finite show that its diagram cannot contain a circuit. [HINT: For a circuit diagram write down a word of arbitrary length which is unique in its homotopy class].

10. If W is finite show that its diagram cannot contain any of the following subdiagrams. [HINT: Apply the previous hint to the first case, and generalize this technique to the other cases].

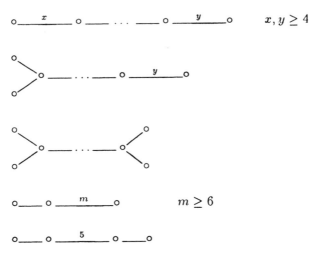

11. (More difficult). If W is finite, then show that its diagram cannot contain any of the following subdiagrams:

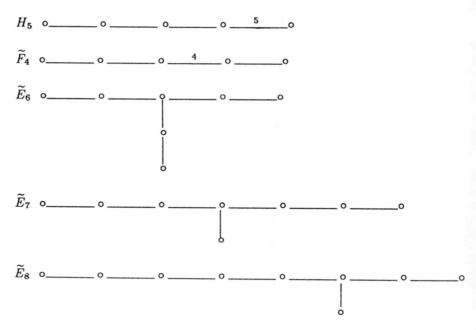

12. Using the results of Exercises 9, 10, and 11 show that if W is finite then its diagram must be the union of connected components, each of which is one of those given in Appendix 5.

Chapter 3
BUILDINGS

This chapter introduces buildings and proves two important properties: the existence of apartments, and the fact that for any chamber c and any residue R, there is a unique chamber of R nearest c. There is also a section on generalized m-gons, which are the same thing as rank 2 buildings.

1. A Definition of Buildings.

We use the notation W, M, I of the previous chapter, and recall that if $f = i_1 \ldots i_k$, then r_f means $r_{i_1} \ldots r_{i_k} \in W$. We can now define a *building of type M*. It is a chamber system Δ over I such that each panel lies on at least two chambers, and having a *W-distance function*

$$\delta : \Delta \times \Delta \to W,$$

such that if f is a reduced word, then $\delta(x, y) = r_f$ if and only if x and y can be joined by a gallery of type f. In particular any two chambers can be joined by a gallery of reduced type. The W-distance $\delta(x, y)$ should not be confused with the distance $d(x, y)$ which is the length of a minimal gallery from x to y; in fact $d(x, y)$ is the length of $\delta(x, y)$ as an element of W. Of course to any building there is an associated cell complex, as in Chapter 1; we shall make no formal distinction between these, and refer to the cells (or simplexes) of a building without further ado.

Example. Coxeter complexes are buildings; simply set $\delta(x, y) = x^{-1}y$.

Remark. If $\gamma = (a, b, c)$ is a gallery of type ii, then either $a = c$ (as for a Coxeter complex) in which case we can replace γ by a null gallery, or else $a \neq c$ in which case we can replace γ by the gallery (a, c) of type i. Thus a gallery of type $f_1 ii f_2$ cannot generally be replaced by one of type $f_1 f_2$. In particular if f is not reduced, the existence of a gallery of type f from x to y does not imply that $\delta(x,y) = r_f$, but on the other hand if $\delta(x,y) = r_f$ then there is a gallery of type f from x to y (Exercise 1).

(3.1). *Here are some elementary consequences of the definition:*

(o) Δ *is connected, δ maps onto W, and $\delta(x,y) = \delta(y,x)^{-1}$.*

(i) $\delta(x,y) = r_i \Leftrightarrow x$ *and y are distinct and i-adjacent.*

(ii) *i- and j-adjacency are mutually exclusive for $i \neq j$.*

(iii) *If there is a gallery of type f (not necessarily reduced) from x to y, and if f is homotopic to g, then there is also a gallery of type g from x to y.*

(iv) *A gallery of type f is minimal $\Leftrightarrow f$ is reduced.*

(v) *If f is reduced, a gallery of type f from x to y is unique.*

PROOF: (o), (i) and (ii) are easy exercises, and (iii) follows from the fact that if there is a gallery of type $p(i,j)$ from x to y, then there is also a gallery of type $p(j,i)$, since both these words are reduced and give the same element of W.

(iv) Let γ be a gallery of type f from x to y. If f is not reduced, then by (iii) we can replace it by a gallery of type $f_1 ii f_2$, and hence a gallery of shorter length, so γ is not minimal. Conversely suppose γ is not minimal, and let g be the type of some minimal gallery. We have shown g is reduced, so if f is also reduced then $r_f = \delta(x,y) = r_g$; therefore $f \simeq g$ by (2.11), contradicting the fact that g is shorter than f.

(v) Let $(x, \ldots, y_1, y)$ and $(x, \ldots, y_2, y)$ be galleries of reduced type fi ($i \in I$) from x to y. Then y_2 is i-adjacent to y_1, because both are i-adjacent to y. Therefore if $y_1 \neq y_2$ we have galleries of reduced types f and fi from x to y_2, a contradiction since $r_f \neq r_{fi}$. Thus $y_1 = y_2$, and a simple induction on the length of the gallery completes the proof. □

2. Generalised m-gons - the rank 2 case.

For any integer $m \geq 2$, or for $m = \infty$, a *generalized m-gon* is a connected, bipartite graph of diameter m and girth $2m$, in which each vertex lies on at least two edges. (A graph is *bipartite* if its set of vertices can be partitioned into two disjoint subsets such that no two vertices in the same subset lie on a common edge; the *diameter* is the maximum distance between two vertices, and the *girth* is the length of a shortest circuit.) If $m = \infty$ this is simply a tree with no end points (Exercise 12).

(3.2) PROPOSITION. *A rank 2 building of type* o——m——o *is a generalized m-gon, and vice versa.*

PROOF: We leave the details to the reader after making two elementary observations. In a Coxeter group of type o——m——o the reduced words are precisely the finite alternating sequences $iji\ldots$ of length $\leq m_{ij}$; they give distinct group elements except for equality between $iji\ldots$ and $jij\ldots$ when both have m_{ij} terms. A generalized m-gon is then considered as a building by taking the edges as chambers, and adjacency to mean having a common vertex, of one of the two appropriate types. □

Example. A generalized 3-gon was illustrated in Figure 1.3 of Chapter 1.

We now define a building to be *thick* if every panel is a face of at least three chambers (i.e. each i-adjacency class has size ≥ 3). It is called *thin* if every panel is a face of exactly two chambers; thin buildings are nothing other than Coxeter complexes, as the reader may immediately verify. The *valency* of a panel will denote the number of chambers having it as a face.

(3.3) PROPOSITION. *In a thick generalized m-gon, vertices of the same type have the same valency, and if m is odd, then all vertices have the same valency.*

PROOF: Define two vertices x and y to be opposite if the distance $d(x, y)$ between them is m; they will be of the same or different type according to whether m is even or odd.

Step 1. Two opposite vertices have the same valency. Given opposite vertices x and y, let e be any edge on x, and let x' be its other vertex. Since x and x' have different types, $d(x', y) < d(x, y)$, and so there is a path from x to y starting with e, and ending with f, say. The girth assumption implies that f is uniquely determined by e, and e by f; this gives a canonical bijection between the set of edges on x and those on y.

Step 2. If x, y are two vertices both joined to a common vertex z, then there exists a vertex opposite both x and y. Indeed since z has valency ≥ 3 we take an edge on z different from zx and zy, and continue this to a path of length $m - 1$ ending at a vertex v. Then $d(x, v) = d(y, v) = m$.

Now if x and y are vertices of the same type we take a path from x to y, and use Steps 1 and 2 to see that x and y have the same valency. If m is odd then opposite vertices have different types, so by Step 1 all vertices have the same valency. □

A generalized m-gon is said to have *parameters* (s,t), where s and t are (possibly infinite) cardinals, if the two valencies are $s+1$ and $t+1$. Before leaving the subject of generalized m-gons, we mention that a generalized 2-gon is simply a complete bipartite graph, and a generalized 3-gon is nothing other than (the flag-graph of) a projective plane. This and other information and examples are contained in the exercises at the end of this chapter. Later on we shall deal with generalized m-gons admitting a large group of automorphisms (the Moufang m-gons); for these important examples $m = 3, 4, 6$ or 8. However there is no such restriction on m in general, as Exercise 21 shows, unless the generalized m-gon is finite. We remark in passing that in this case W. Feit and G. Higman [1964] proved the following theorem using character theory. We shall not prove it, but simply refer the reader to [loc. cit.], and also to D. Higman [1975].

(3.4) THEOREM. *(W. Feit - G. Higman): A finite thick generalized m-gon exists only if $m = 2, 3, 4, 6$ or 8. Moreover if the parameters are (s,t) then there are restrictions on s and t such as:*

$$\text{for } m = 4 \quad \frac{st(st+1)}{s+t} \in \mathbf{Z}$$
$$\text{for } m = 6 \quad st \text{ is a perfect square}$$
$$\text{for } m = 8 \quad 2st \text{ is a perfect square}$$

Moreover, D. Higman [1975] proves that for $m = 4$ or 8, $s \le t^2$ and $t \le s^2$ (see also Exercise 19); and W. Haemers [1979] proves that for $m = 6$, $s \le t^3$ and $t \le s^3$.

3. Residues and Apartments.

We now continue with further basic results on buildings. If J is a subset of I, then as in Chapter 2, M_J is the subdiagram spanned by the elements of J (i.e., all m_{ij}, for $i, j \in J$), and W_J the appropriate Coxeter group (cf.(2.14)). For the rest of this chapter Δ will denote a building of type M.

(3.5) THEOREM. *Every J-residue of Δ is a building of type M_J.*

PROOF: It suffices to show that if x and y are any two chambers in a common J-residue then $\delta(x,y) \in W_J$, so let γ be a shortest J-gallery joining them. If its type f is not reduced, then by (3.1)(iii) there is a J-gallery of type $f_1 iif_2$ from x to y, and hence a shorter J-gallery. Thus f is reduced, and $\delta(x,y) = r_f \in W_J$. $\square$

Given any subset $X \subset W$ we define a map $\alpha : X \to \Delta$ to be an *isometry* if it preserves the W-distance δ. In other words, using δ_W for distance in W, and δ_Δ for distance in Δ, we require

$$\delta_\Delta(\alpha(x), \alpha(y)) = \delta_W(x, y)$$

for all $x, y \in X$; recall that $\delta_W(x, y) = x^{-1}y$.

An *apartment* will mean an isometric image $\alpha(W)$ of W in Δ. A *root* or *wall* of Δ will mean a root or wall in an apartment of Δ; notice that if X is a root (or wall) in an apartment A, then the same is true for any apartment containing X. Moreover by the following theorem an isometric image of a root of W is a root of Δ.

(3.6) THEOREM. *Any isometry of a subset $X \subset W$ into Δ extends to an isometry of W into Δ.*

PROOF: Let $\alpha : X \to \Delta$ denote the isometry, and assume $X \neq W$. By Zorn's lemma it suffices to extend the domain of α to a strictly larger subset of W. If $X = \emptyset$ this is a triviality, so suppose X is non-empty, in which case we can find $x_o \in X$ and $i \in I$ such that $x_o r_i \notin X$. Modifying X and α by $x_o^{-1} \in W$, we may assume $x_o = 1 \in X$ and $r_i \notin X$. We extend α by defining $\alpha(r_i)$.

Case 1. $\ell(r_i x) > \ell(x)$ for all $x \in X$ (Figure 3.1)

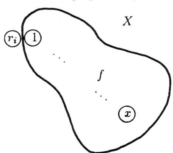

Figure 3.1

In this case let $\alpha(r_i)$ be any chamber distinct from and i-adjacent to $\alpha(1)$. We need to show that $\delta(\alpha(r_i), \alpha(x)) = r_i x$ for all $x \in X$, so let $x = r_g$ with g reduced. Then there is a gallery of type ig from $\alpha(r_i)$ to $\alpha(x)$, and since $\ell(r_i x) > \ell(x)$ we know ig to be reduced. Thus $\delta(\alpha(r_i), \alpha(x)) = r_{ig} = r_i r_g = r_i x$, as required.

Case 2. $\ell(r_i x_1) < \ell(x_1)$ for some $x_1 \in X$. In this case there is, in W, a minimal gallery from 1 to x_1 via r_i of reduced type f. Let y be the second term in the unique gallery of type f from $\alpha(1)$ to $\alpha(x_1)$, and define $\alpha(r_i) = y$. Again we need to show that $\delta(y, \alpha(x)) = r_i x$ for all $x \in X$, so define $\beta(x) = r_i \delta(y, \alpha(x))$. Since y is i-adjacent to $\alpha(1)$, we see that $\delta(y, \alpha(x)) = r_i x$ or x, and therefore

$$\beta(x) = x \text{ or } r_i x.$$

Now, as a map from X to W, β is a composite of three maps: α, $\delta(y, \ )$ and r_i (left multiplication). The first and last of these preserve distances, and the middle one does not increase distances, because it preserves adjacency. Therefore β does not increase distances, and moreover $\beta(1) = 1$ and $\beta(x_1) = x_1$. Now if α_i is the root of W containing 1 but not r_i (see Chapter 2), then $x_1 \in -\alpha_i$. Therefore $\beta(x) \neq r_i x$ otherwise β increases either the distance from 1 to x (if $x \in \alpha_i$) or the distance from x_1 to x (if $x \in -\alpha_i$) because in each case $r_i x$ lies in the opposite root. This contradiction shows that $\beta : X \to W$ is the inclusion map, and hence $\delta(y, \alpha(x)) = r_i x$. □

(3.7) COROLLARY. *Any two chambers lie in a common apartment.* □

Notice that an isometry $\alpha : W \to \Delta$ is uniquely determined by its image $A = \alpha(W)$ together with the chamber $c = \alpha(1)$, because if α' is another such isometry, then $\alpha^{-1} \alpha'$ is an isometry of W fixing the element $1 \in W$, and is therefore the identity map. Now fix any apartment A and chamber $c \in A$. We define a map

$$\rho_{c,A} : \Delta \to A$$

called the *retraction* of Δ *onto* A with *centre* c. Let $A = \alpha(W)$ with $\alpha(1) = c$, and set

$$\rho_{c,A}(x) = \alpha(\delta(c, x)).$$

It is straightforward to see that for $x \in A$, $\rho_{c,A}(x) = x$; and indeed as a map of simplicial complexes $\rho_{c,A}$ is a retraction in the usual topological sense.

If σ and τ are simplexes, a gallery *from σ to τ* means a gallery $(c, \dots, d)$ where σ is a face of c, and τ a face of d; of course if σ is a chamber then $c = \sigma$.

(3.8) THEOREM. *Let A be an apartment containing a chamber c and a simplex σ. Then every minimal gallery from c to σ lies in A; in particular apartments are convex.*

PROOF: Let $\gamma = (c = c_o, c_1, \ldots, c_k)$ be a minimal gallery from c to σ. If $\gamma \not\subset A$ then for some t, $c_{t-1} \in A$ and $c_t \notin A$. Let $b \neq c_{t-1}$ be the other chamber of A adjacent to c_{t-1} and c_t, so $\rho_{b,A}(c_{t-1}) = \rho_{b,A}(c_t)$. Hence $\rho_{b,A}(\gamma)$ contains a repetition and therefore gives a shorter gallery from c to σ, contradicting the minimality of γ. □

(3.9) COROLLARY. *If σ is any simplex of Δ (i.e., $St(\sigma)$ is any residue), and c is any chamber, then there is a unique chamber nearest c having σ as a face (i.e., belonging to $St(\sigma)$).*

PROOF: By (3.7) c and σ (in fact c and any chamber having σ as a face) lie in a common apartment A. By (3.8) any chamber having σ as a face and at minimal distance from c lies in A. The result now follows from the same result for W, namely (2.9). □

The chamber of $St(\sigma)$ nearest c in (3.9) will be called $\mathrm{proj}_\sigma c$, or $\mathrm{proj}_R c$ if $R = St(\sigma)$.

Direct Products and Disconnected Diagrams. Let $M = M_1 \cup \ldots \cup M_k$ be the decomposition of the diagram into connected components, where M_t is over the set I_t. In particular I is the disjoint union $I_1 \cup \ldots \cup I_k$, and $m_{ij} = 2$ if i and j belong to different components. Fix some chamber c of a building Δ of type M, and let Δ_t denote the I_t-residue containing c.

(3.10) THEOREM. *With the notation above, Δ is isomorphic to the direct product $\Delta_1 \times \ldots \times \Delta_k$.*

PROOF: Setting $W_t = W_{I_t}$ (the Coxeter group of type M_t), we have $W = W_1 \times \ldots \times W_k$ by Exercise 7 of Chapter 2, and so any $w \in W$ can be written $w_1 \ldots w_k$ where $w_t \in W_t$, and for each t we may write $w = w_t w_t'$, where $w_t' = w_1 \ldots \hat{w}_t \ldots w_k$ (w_t removed).

Now let $d \in \Delta$ be any chamber, let $w = \delta(c, d)$, and let d_t denote the unique chamber at distance w_t from c on a minimal gallery from c to d, characterised by

$$\delta(c, d_t) = w_t \text{ and } \delta(d_t, d) = w_t'.$$

We define a map $\varphi : \Delta \rightarrow \Delta_1 \times \ldots \times \Delta_k$ via

$$\varphi(d) = (d_1, \ldots, d_k).$$

and show it to be an isomorphism. If R is an I_t-residue, then φ followed by projection to Δ_t maps R isomorphically onto Δ_t (indeed if γ_0 is a gallery of reduced type f_0 from c to $\text{proj}_R c$, and γ a gallery of reduced type f in R from $\text{proj}_R c$ to d, then there is a unique gallery $\gamma'\gamma_0'$ of type ff_0 from c to d, and $\varphi(\gamma) = \gamma'$). This shows φ is a surjective morphism. To show injectivity, suppose $\varphi(d) = \varphi(d')$, so in particular $\delta(c,d) = \delta(c,d') = w = w_1 \ldots w_k$. Take galleries $\gamma = \gamma_1 \ldots \gamma_k$ and $\gamma' = \gamma_1' \ldots \gamma_k'$ from c to d and from c to d' respectively, where γ_t and γ_t' are I_t-galleries. Obviously γ_1 and γ_1' have the same end chamber d_1, and so γ_2 and γ_2' are galleries in the same I_2-residue. Since $\varphi(\gamma_2)$ and $\varphi(\gamma_2')$ have the same end chamber d_2, so do γ_2 and γ_2'. An obvious induction shows $d = d'$. $\square$

An Alternative Definition. The definition given at the beginning of this chapter is of recent vintage. Earlier definitions presupposed the existence of apartments in some form or other, and we now give a formulation of this sort. It can be used to check that a given chamber system is a building, without needing to define a W-distance having the required properties (cf. Exercise 8).

(3.11) THEOREM. *Let C be a chamber system containing subsystems (called apartments) isomorphic to a given Coxeter complex (over the same indexing set I), and such that any two chambers lie in a common apartment. Then C is a building if, given two apartments A and A' containing a common chamber x and chamber or panel y, A and A' are isomorphic via an isomorphism fixing x and y.*

PROOF: Given chambers x and y we define $\delta(x,y)$ to be the W-distance in any apartment containing x and y; by hypothesis this is well-defined. Furthermore if f is a reduced word and $\delta(x,y) = r_f$, then there is a gallery of type f from x to y in any such apartment. Conversely assume there is a gallery $(x, \ldots, y', y)$ of reduced type $f = gi$ $(i \in I)$ from x to y; then we must show that $\delta(x,y) = r_f$. Let A be an apartment containing x and y, and let π be the panel (of type i) common to y and y'. By induction on the length of f, we know that $\delta(x,y') = r_g$, and therefore there is a gallery γ of type g in an apartment A', from x to y'. Let $\varphi : A' \rightarrow A$ be an isomorphism fixing x and π. Then $(\varphi(\gamma), y)$ is a gallery of type $gi = f$ in A from x to y, so $\delta(x,y) = r_f$, as required. $\square$

Notes. The definition of a building at the beginning of this chapter is given in Tits [1986b]. It is equivalent to the definition given by Tits [1974] which is much closer to that furnished by Theorem (3.11). The proof of Theorem (3.6) is taken from Tits [1981], where chamber systems were first introduced.

Exercises to Chapter 3

1. If $\delta(x,y) = r_f$ with f not necessarily reduced, show there is a gallery of type f from x to y.

2. If A is any apartment and σ a simplex in A, show that $A \cap St(\sigma)$ is an apartment of $St(\sigma)$.

3. Let α be a root, and π a panel in $\partial\alpha$. If $x, y \notin \alpha$ are chambers in $St(\pi)$, show that $\alpha \cup \{x\}$ is isometric to $\alpha \cup \{y\}$. Conclude that $\alpha \cup \{x\}$ lies in an apartment, and show that α is the intersection of all apartments containing it.

4. Given any two chambers x and y in a thick building, show that the set of all chambers on minimal galleries from x to y is the same as the intersection of all apartments containing both x and y. [HINT: Use (3.8), (2.8) and Exercise 3].

5. Let W be finite, and define two chambers x and y to be *opposite* if they are opposite in some apartment A containing both. Show that A is the only apartment containing both x and y. [HINT: Use (2.5) (iv) and Exercise 4].

6. Let A and A' be apartments having a chamber in common. Show that $A \cap A'$ is a convex set of chambers (together with their faces), and that there is an isomorphism from A to A' fixing $A \cap A'$. [HINT: Use (3.6)].

7. Let A and A' be apartments containing simplexes σ and τ. Show there is an isomorphism from A to A' fixing σ and τ. [HINT: Take chambers $c \in St(\sigma) \cap A$ and $d \in St(\tau) \cap A'$, and let A'' be an apartment containing c and d].

8. Show that Example 4 of Chapter 1 is a building. [HINT: Use (3.11)].

9. In Example 4 of Chapter 1, let $c = (V_1 \subset V_2 \subset \ldots \subset V_n)$ be any chamber, and let $\sigma = W_n$ be any subspace of dimension n. Find the unique chamber nearest c having σ as a face, as in (3.9).

10. Let Δ be a building and let Δ' be a sub-chamber system which is a union of apartments such that any two chambers of Δ' lie in one of these apartments. Show that Δ' is a building (having the same type as Δ, of course).

11. If $R_1, \ldots, R_t$ are residues of types $J_1, \ldots, J_t$ in a building Δ, show that $R_1 \cap \ldots \cap R_t$ is a residue of type $J_1 \cap \ldots \cap J_t$, and hence if Δ has finite rank its geometric realisation (in the sense of Chapter 1 section 1) is a simplicial complex. [HINT: Use (2.1)(ii)].

12. Show that a generalized ∞-gon (i.e., W infinite dihedral) is the same thing as a tree with no end point (i.e. no vertex on only a single edge).

13. Show that a generalized 2-gon is a complete bipartite graph (i.e., two sets of vertices X and Y with edges being all pairs $\{x, y\}$ with $x \in X$, $y \in Y$).

14. Given a generalized m-gon Δ with $m \geq 3$, call the two types of vertices *points* and *lines* and define a point to be *on* a line if they are the vertices of a common edge of Δ. Using this interpretation, show that thick generalized 3-gons are the same thing as projective planes (i.e., any two distinct points lie on a unique common line, any two lines have a point in common, and there exists a non-degenerate quadrangle.)

15. Given parameters (s, t) for a generalized m-gon with m finite, show that the number of chambers (edges) opposite a given chamber is $(st)^{m/2}$ if m is even, and $s^{\frac{m+1}{2}} t^{\frac{m-1}{2}}$ if m is odd; for m odd, reversing the roles of s and t gives an alternative proof that $s = t$. If m is even show that the total number of chambers is $(s + 1)(t + 1)(1 + st + \ldots + (st)^{\frac{m}{2}-1})$.

16. Let Δ be a generalized $2m$-gon having vertices of types 1 and 2, and suppose each vertex of type 1 has valency 2 (i.e. lies on exactly 2 edges). Show that Δ is obtained from a generalized m-gon Δ_0 by introducing a new vertex (of type 1 in Δ) in the middle of each edge of Δ_0 and taking the vertices of Δ_0 to be the type 2 vertices of Δ. If Δ has parameters $(1, t)$ with $t \neq 1$, conclude that Δ_0 has parameters (t, t), and hence by the Feit-Higman Theorem (3.4) that $m = 2, 3, 4$ or 6 only.

17. A *polarity* of a generalized m-gon is an (outer) automorphism of order 2 interchanging the two types of vertices. Show that the chambers fixed by a polarity are mutually opposite, and if there are no fixed chambers then every chamber is carried to an opposite one.

18. Show that generalized 4-gons (quadrangles) are those point-line geometries, in the sense of Exercise 14, satisfying:

   (i) two points lie on at most one line;

   (ii) there exists a non-degenerate quadrangle;

(iii) for any line L and point p not on L, there is a unique point of L collinear with p.

19. Let V be a 4-dimensional vector space over k with basis $\{x_1, x_2, y_1, y_2\}$ and alternating bilinear form

$$(x_i, x_j) = (y_i, y_j) = 0$$

and

$$(x_i, y_j) = -(y_j, x_i) = \delta_{ij}.$$

Let *points* be 1-spaces and *lines* be totally isotropic 2-spaces S (i.e., $(s,t) = 0 \ \forall s, t \in S$). Show that this is a generalized quadrangle in the sense of Exercise 18. If $k = F_q$, it has parameters (q, q).

20. Let Q denote the geometry of Exercise 19, and let p be any point (1-space) of Q. Define a new geometry Q' as follows:

points of Q' are points of Q not collinear with p;

lines of Q' are all lines of Q not on p, and all non-isotropic 2-spaces containing p.

Show that, with the obvious incidence (containment) relation, Q' is a generalized quadrangle. If $k = F_q$ it has parameters $(q - 1, q + 1)$.

21. Let G be any bipartite graph of finite diameter and girth $2m$ containing a circuit. Adjoin new vertices and edges to get a larger graph G' as follows: if x and y are vertices of G with $d(x,y) = m + 1$, introduce $m - 1$ new edges and $m - 2$ new vertices forming a chain of length $m - 1$ joining x to y. Then G' has girth $2m$, finite diameter, and contains a circuit. Moreover if $x, y \in G$ and $d_G(x,y) = d > m$, then $d_{G'}(x,y) < d$. Repeating this procedure ad infinitum, show that one obtains a generalized m-gon $\overline{G}$. What modification is necessary to ensure that $\overline{G}$ is thick?

22. (P.J.Cameron) Consider a generalized quadrangle with parameters (s, t). For any point x, let $x^\perp$ denote the set of points collinear with x. Two points x and y which are not collinear are called *opposite*; in this case $|x^\perp \cap y^\perp| = t + 1$. Now let $\{z_1, \dots, z_n\}$ be the set of points opposite both x and y, and for each z_i let $a_i = |x^\perp \cap y^\perp \cap z_i^\perp|$.

(i) Show that $n = s^2 t - st - s + t$.

(ii) Show that $\Sigma a_i = (t + 1)(t - 1)s$ and $\Sigma a_i(a_i - 1) = (t + 1)t(t - 1)$.
[HINT: Count pairs (v, z_i) and triples (v, w, z_i) where $v, w \in x^\perp \cap y^\perp \cap z_i^\perp$].

(iii) Using the inequality $(\Sigma a_i)^2 \leq n\Sigma a_i^2$, derive the inequality $(s-1)(s^2 - t) \geq 0$.

(iv) Conclude that for a thick generalized quadrangle, $t \leq s^2$ and (dually) $s \leq t^2$.

 (v) If $t = s^2$ what does this say about the number of points collinear with three mutually opposite points?

Chapter 4
LOCAL PROPERTIES AND COVERINGS

This chapter deals mainly with coverings of chamber systems (defined in section 2), particularly those chamber systems whose rank 2 residues are buildings. Most of the chapter is independent of the rest of this book; in particular there is no connection with Chapters 5 and 6, and only section 1 will be used in Chapter 7.

1. Chamber Systems of Type M.

In Theorem 3.5 we saw that every residue of a building is a building; in particular by (3.2), the $\{i,j\}$-residues are generalised m_{ij}-gons. We now define a *chamber system of type M* to be a chamber system over I for which each $\{i,j\}$-residue is a generalised m_{ij}-gon. Although such a chamber system is not necessarily a building, we shall show that its universal cover (section 3) is a building, provided the same is true for all J-residues whenever $|J| = 3$ and W_J is finite.

In a chamber system of type M, we define a *strict elementary homotopy* of galleries to be an alteration from a gallery of the form $\gamma_1 \gamma_0 \gamma_2$ to one of the form $\gamma_1 \gamma_0' \gamma_2$ where γ_0 has type $p(i,j)$ and γ_0' has type $p(j,i)$. Two galleries are then called *strictly homotopic* if one can be transformed into the other via a sequence of strict elementary homotopies.

Notice that if γ and γ' are strictly homotopic galleries of types f and f', then f and f' are homotopic as words and hence $r_f = r_{f'}$; thus each strict homotopy class of galleries determines an element of W.

(4.1) LEMMA. *Let C be a chamber system of type M. Given a gallery γ in C of reduced type f from x to y, and a homotopy $f \simeq g$ of words, there* *exists a gallery γ' of type g from x to y which is strictly homotopic to γ. Moreover a minimal gallery must have reduced type.*

PROOF: In a generalised m_{ij}-gon a gallery of type $p(i, j)$ is certainly strictly homotopic to one of type $p(j, i)$ so a homotopy of words may be realised at the gallery level, proving the first statement. To prove the second statement, let γ be a minimal gallery of type f from x to y. If f is not reduced it is homotopic to a word of the form $f_1 i i f_2$, and hence there is a shorter gallery, of type $f_1 i f_2$ or $f_1 f_2$, from x to y. $\qquad\square$

We now give a characterization of buildings as connected chamber systems of type M satisfying the following condition for one single chamber.

(P_x). *If two reduced words f, f' are the types of two galleries from x to some common chamber, then $r_f = r_{f'}$.*

If (P_x) is satisfied, then there is a well-defined distance $\delta(x, y) = r_f$ from x to any other chamber y, where f is a reduced word which is the type of a gallery from x to y (such a gallery exists and is obviously minimal, cf. 4.1). Moreover if $r_f = r_g$ (g reduced), then $f \simeq g$ by (2.11), and by (4.1) there is also a gallery of type g from x to y. Thus if (P_x) is satisfied, $\delta(x, y) = r_f$ (f reduced) if and only if there is a gallery of type f from x to y.

(4.2) THEOREM. *A connected chamber system C of type M is a building if and only if (P_c) holds for some chamber $c \in C$.*

PROOF: By definition (P_c) holds for all chambers in a building. Conversely the preceding discussion shows that C is a building if (P_x) holds for all x. By connectivity it therefore suffices to prove that $(P_c) \Rightarrow (P_{c'})$ whenever c' is adjacent to c.

We suppose c' is j-adjacent to c, and $c' \neq c$. Given two galleries γ, γ' from c' to d, having reduced types f, f' we must show that $r_f = r_{f'}$.

Case 1. Suppose both jf and jf' are reduced.

By applying (P_c) to the galleries (c, γ) and (c, γ') one has $r_{jf} = r_{jf'}$, and hence $r_f = r_{f'}$.

Case 2. Neither jf nor jf' is reduced.

By (2.13) $f \simeq jg$ and $f' \simeq jg'$ where both jg and jg' are reduced. By (4.1) we therefore have galleries $(c', \gamma_1) = (c', c_1, \dots, d)$ and $(c', \gamma'_1) = (c', c'_1, \dots, d)$ of types jg and jg' respectively. Clearly c, c', c_1 and c'_1 are all mutually j-adjacent.

If $c_1 = c = c_1'$ we apply (P_c) to γ_1 and γ_1' to conclude that $r_g = r_{g'}$ and hence $r_f = r_{f'}$.

If $c_1 \neq c \neq c'$ we apply (P_c) to (c, γ_1) and (c, γ_1') to conclude that $r_{jg} = r_{jg'}$, and hence $r_f = r_{f'}$.

If $c_1 = c \neq c_1'$ we apply (P_c) to (γ_1) and (c, γ_1') to conclude that $r_g = r_{jg'}$. This implies $g \simeq jg'$, contradicting the fact that jg is reduced. A similar contradiction eliminates the possibility $c_1 \neq c = c_1'$, completing the proof of Case 2.

Case 3. Exactly one of jf or jf' is reduced.

We show this cannot happen. Without loss of generality jf is reduced and jf' is not, so by (2.13) $f' \simeq jg$. As in Case 2 we have a gallery $(c', \gamma_1) = (c', c_1, \ldots, d)$ of type jg, and c, c' and c_1 are mutually j-adjacent.

If $c = c_1$ we apply (P_c) to γ_1 and (c, γ) to conclude that $r_g = r_{jf}$, hence $g \simeq jf$, so $f' \simeq jg$ is not reduced, a contradiction.

If $c \neq c_1$ we apply (P_c) to (c, γ_1) and (c, γ) to conclude that $r_{jg} = r_{jf}$, and hence $g \simeq f$. Therefore by (4.1) there is a gallery γ' of type f from c_1 to d. The galleries (c, γ) and (c, γ') both have reduced type jf; this implies, using (P_c), that they must be the same gallery (the proof of (3.1)(v) goes through unchanged), and hence $c' = c_1$, a contradiction. □

2. Coverings and the Fundamental Group.

A morphism $\varphi : C \to D$ of chamber systems is called a *covering* if it maps each rank 2 residue of C isomorphically onto a rank 2 residue of D of the same type (the term 2-*covering* is also used). We say also that C *covers* D.

Remark on Topology. Any chamber system C of finite rank n has a geometric realization as a CW-complex Δ of dimension $n - 1$, built from simplexes, as explained in Chapter 1, section 1. If $\varphi_\Delta : \widetilde{\Delta} \to \Delta$ is a covering of topological spaces, then $\widetilde{\Delta}$ inherits a cellular decomposition from Δ, and can be viewed as the geometric realization of a chamber system $\widetilde{C}$ (chambers being faces of dimension $n - 1$, panels being faces of dimension $n - 2$). Since φ_Δ is a homeomorphism in the neighborhood of each point, it induces a map $\varphi : \widetilde{C} \to C$ which is an isomorphism on each residue of rank $< n$. We shall call such coverings *topological*; if $n \geq 3$ every topological covering is a covering in the sense defined above, and for $n = 3$ the two concepts coincide. Of course for $n > 3$ our coverings need not be isomorphisms on rank 3 residues; in topological terms they are "branched" (or "ramified") over a subcomplex of codimension ≥ 3.

To investigate coverings of topological spaces one uses the "fundamental group" whose elements are homotopy classes of paths beginning and ending at some given vertex. There is an analogous notion for chamber systems, which we now discuss.

In any chamber system an *elementary homotopy* of galleries is an alteration from a gallery of the form $\gamma\omega\delta$ to $\gamma\omega'\delta$ where ω and ω' are galleries (with the same extremities) in a rank 2 residue. We then say that two galleries are *homotopic* if one can be transformed to the other by a sequence of elementary homotopies. Notice that in a chamber system of type M two galleries which are strictly homotopic are obviously homotopic.

If c is a chamber in a connected chamber system C, a *closed gallery based at c* will mean a gallery starting and ending at c. The *fundamental group* $\pi(C,c)$ is the set of homotopy classes $[\gamma]$ of closed galleries γ based at c, together with the binary operation $[\gamma] \cdot [\gamma'] = [\gamma\gamma']$ where $\gamma\gamma'$ means γ followed by γ'; using γ^{-1} to denote the reversal of γ, one has $[\gamma]^{-1} = [\gamma^{-1}]$.

Notice that if c' is any other chamber, and δ is a gallery from c to c', then $[\gamma] \to [\delta^{-1}\gamma\delta]$ gives an isomorphism from $\pi(C,c)$ to $\pi(C,c')$. We call C *simply-connected* if it is connected and $\pi(C,c) = 1$. Given a morphism $\varphi : C \to D$ with $\varphi(c) = d$ (sometimes written $\varphi : (C,c) \to (D,d)$), one defines a map

$$\varphi_* : \pi(C,c) \to \pi(D,d)$$

via $[\gamma] \to [\varphi(\gamma)]$; this is obviously a group homomorphism, and if φ is a covering it is injective (Exercise 1).

(4.3) THEOREM. *Buildings are simply-connected.*

PROOF: Let γ be any closed gallery based at c which is minimal in its homotopy class. If $\gamma \neq (c)$ then its type f is not reduced, otherwise $\delta(c,c) = r_f$; therefore there is a sequence of elementary homotopies from f to a word of the form $f_1 i i f_2$. By (4.1) γ is strictly homotopic to a gallery of this type and therefore homotopic to a shorter gallery, of type $f_1 i f_2$ or $f_1 f_2$, a contradiction. Thus the fundamental group is trivial. □

(4.4) LEMMA. *Let $\varphi : C \to D$ be a covering. Given a gallery γ in D starting at some chamber x, and given $\tilde{x} \in \varphi^{-1}(x)$, there is a unique gallery $\tilde{\gamma}$ in C starting at $\tilde{x}$ and with $\varphi(\tilde{\gamma}) = \gamma$.*

PROOF: Exercise. □

(4.5) LEMMA. *Given coverings* $\varphi : (C, c) \rightarrow (D, d)$ *and* $\psi : (E, e) \rightarrow (D, d)$ *with* C *and* E *connected, there exists a covering* $\alpha : (C, c) \rightarrow (E, e)$ *with* $\psi\alpha = \varphi$ *if and only if* $\varphi_* \pi(C, c) \leq \psi_* \pi(E, e)$.

$$
\begin{array}{ccc}
C & \xrightarrow{\;\alpha\;} & E \\
& \varphi \searrow \qquad \swarrow \psi & \\
& D &
\end{array}
$$

PROOF: If α exists, then for any $[\gamma] \in \pi(C, c)$ we have

$$\varphi_*[\gamma] = (\psi\alpha)_*[\gamma] = [\psi\alpha(\gamma)] = \psi_*[\alpha(\gamma)] \in \psi_*\pi(E, e).$$

Conversely, to define α, take any chamber x of C, and let γ be a gallery in C from c to x. By (4.4) the gallery $\varphi(\gamma)$ in D has a unique lifting to a gallery ϵ in E starting at e. The final chamber of ϵ is defined to be $\alpha(x)$, (see Figure 4.1); obviously $\psi\alpha(x)$ is the final chamber of $\varphi(\gamma)$, namely $\varphi(x)$, and hence $\psi\alpha = \varphi$. We must show α is well-defined; it will then follow that α is a morphism, and hence a covering since ψ and φ are.

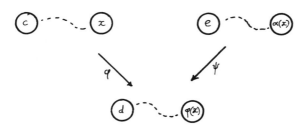

Figure 4.1

Thus let γ' be another gallery from c to x, and let ϵ' be the lifting of $\varphi(\gamma')$ starting at e. By hypothesis $\varphi(\gamma'\gamma^{-1})$ is homotopic to $\psi(\delta)$ for some closed gallery δ in E based at e. Any sequence of elementary homotopies from $\psi(\delta)$ to $\varphi(\gamma'\gamma^{-1})$ lifts to E giving a homotopy from δ to a closed gallery θ with $\psi(\theta) = \varphi(\gamma'\gamma^{-1})$. By the uniqueness of liftings (4.4) one has $\theta = \epsilon'\epsilon^{-1}$; thus ϵ' has the same end chamber as ϵ, and α is well-defined. $\square$

Remark. If g is an automorphism of C, then $g \circ \varphi : (\widetilde{C}, \tilde{c}) \to (C, c)$ is a covering, and using (4.5) one finds that g lifts to an automorphism $\tilde{g}$ of $\widetilde{C}$ sending $\tilde{c}$ to $\widetilde{g(c)} \in \varphi^{-1}(g(c))$ if and only if $g_* \varphi_* \pi(\widetilde{C}, \tilde{c}) = \varphi_* \pi(\widetilde{C}, \widetilde{g(c)})$ (Exercise 5).

$$(\widetilde{C}, \tilde{c}) \quad \xrightarrow{\ g\ } \quad (\widetilde{C}, \widetilde{g(c)})$$

$$g \circ \varphi \searrow \qquad \swarrow \varphi$$

$$(C, g(c))$$

Since g_* is an automorphism of $\pi(C, c)$, this will certainly be the case whenever $\varphi_* \pi(\widetilde{C}, \tilde{c})$ is a characteristic subgroup of $\pi(C, c)$; in particular whenever $\pi(\widetilde{C}, \tilde{c}) = 1$ (see Exercises 6-8).

3. The Universal Cover.

We assume from here on that all our chamber systems are connected.

Definition. A covering $\varphi : (\widetilde{C}, \tilde{c}) \to (C, c)$ is called *universal* if whenever $\psi : (\overline{C}, \overline{c}) \to (C, c)$ is a covering there exists some covering $\alpha : (\widetilde{C}, \tilde{c}) \to (\overline{C}, \overline{c})$ such that $\psi \alpha = \varphi$.

(4.6) PROPOSITION. *Universal coverings always exist and are unique up to isomorphism. Moreover a covering $\varphi : (\widetilde{C}, \tilde{c}) \to (C, c)$ is universal if and only if $\widetilde{C}$ is simply-connected (i.e., $\pi(\widetilde{C}, \tilde{c}) = 1$).*

PROOF: Uniqueness up to isomorphism follows from the universal property as usual. Moreover if $\pi(\widetilde{C}, \tilde{c}) = 1$, then (4.5) implies that $\widetilde{C}$ is universal. To prove the converse it suffices to construct a simply-connected covering $\widetilde{C}$, as follows.

The chambers of $\widetilde{C}$ are homotopy classes of galleries in C starting at c, and we let $\tilde{c}$ denote the class of the trivial gallery $[c]$. Define i-adjacency by $[c_0, \ldots, c_{k-1}, c_k] \underset{i}{\sim} [c_0, \ldots, c_{k-1}, c_k']$ and $[c_0, \ldots, c_{k-1}]$ where $c_k' \underset{i}{\sim} c_k$, and define φ by $\varphi[c, \ldots, d] = d$. In a rank 2 residue two galleries are homotopic if and only if they have the same end chambers, so φ is an isomorphism when restricted to rank 2 residues. To see that $\pi(\widetilde{C}, \tilde{c}) = 1$ let $\tilde{\gamma}$ be a closed gallery in $\widetilde{C}$ based at $\tilde{c}$. The definition of $\widetilde{C}$ implies that when we lift a gallery δ of C starting at c, to a gallery $\tilde{\delta}$ of $\widetilde{C}$ starting at $\tilde{c}$, the end chamber of $\tilde{\delta}$ is the homotopy class of δ. Since $\tilde{\gamma}$ has end chamber $\tilde{c}$, the homotopy class of $\varphi(\tilde{\gamma})$ is that of the null gallery (c). Moreover, since φ is an isomorphism when restricted to rank 2 residues, each elementary homotopy in C can be lifted to $\widetilde{C}$, and therefore a homotopy in C from

$\varphi(\tilde{\gamma})$ to (c) lifts to a homotopy in $\tilde{C}$ from $\tilde{\gamma}$ to the null gallery $(\tilde{c})$, showing $\pi(\tilde{C}, \tilde{c}) = 1$.   □

(4.7) PROPOSITION. *Let* $\varphi : (\tilde{C}, \tilde{c}) \rightarrow (C, c)$ *be a universal covering of a chamber system of type* M. *Then* $\tilde{C}$ *is a building if and only if whenever two galleries in* C *starting at* c, *and of reduced type, are homotopic they are strictly homotopic.*

PROOF: Suppose homotopic implies strictly homotopic; to show $\tilde{C}$ is a building it suffices, by (4.2), to verify $P_{\tilde{c}}$. So let γ, γ' be two galleries in $\tilde{C}$ from $\tilde{c}$ to some common chamber and of reduced types f, f' respectively. Since $\tilde{C}$ is simply-connected γ and γ' are homotopic. Therefore $\varphi(\gamma)$ and $\varphi(\gamma')$ are also homotopic, so by hypothesis there is a sequence of strict elementary homotopies from $\varphi(\gamma)$ to $\varphi(\gamma')$, and these pull back under φ^{-1} to show that γ and γ' are strictly homotopic. Therefore $f \simeq f'$, and $r_f = r_{f'}$ as required.

Now suppose $\tilde{C}$ is a building and let γ_1, γ_2 be galleries in C of reduced types f_1, f_2 starting at c, and which are homotopic. The unique liftings (see 4.4) to galleries $\tilde{\gamma}_1, \tilde{\gamma}_2$ in $\tilde{C}$ starting at $\tilde{c}$ therefore have the same end chamber. Since f_1 and f_2 are reduced we have $f_1 \simeq f_2$ and therefore by (4.1) $\tilde{\gamma}_1$ is strictly homotopic to a gallery of type f_2 and by uniqueness (3.1)(v) this is $\tilde{\gamma}_2$. The appropriate sequence of strict elementary homotopies is mapped by φ to a sequence of strict elementary homotopies from γ_1 to γ_2, showing γ_1 and γ_2 are strictly homotopic as required.   □

(4.8) PROPOSITION. *The universal cover of a chamber system* C *of type* M *is a building if and only if* (R_c) *holds for some* $c \in C$.

(R_c). *Any two galleries from* c *to a common chamber which are strictly homotopic and of the same reduced type must be equal.*

PROOF: Let $\varphi : (\tilde{C}, \tilde{c}) \rightarrow (C, c)$ be a universal cover. If γ_1, γ_2 are galleries in C of reduced type f starting at c, and which are strictly homotopic, then their liftings $\tilde{\gamma}_1, \tilde{\gamma}_2$ to galleries in $\tilde{C}$ starting at $\tilde{c}$ have the same end chambers and the same type f. By (3.1)(v), $\tilde{\gamma}_1 = \tilde{\gamma}_2$, hence $\gamma_1 = \gamma_2$, and (R_c) holds.

Conversely suppose (R_c) is satisfied. We shall show that homotopy implies strict homotopy, in the sense of (4.7), and hence $\tilde{C}$ is a building. Recall first that, as in (4.5), the chambers of $\tilde{C}$ correspond to homotopy classes of galleries in C starting at c, and $\varphi[c, \ldots, d] = d$. Similarly we define $\overline{C}$ by taking its chambers to be strict homotopy classes of galleries of reduced type in C starting at c. We let $[\gamma]_s$ denote the strict homotopy class of γ, and let $\bar{c} = [c]_s$, (the class of the null-gallery). Adjacency is defined as in (4.5) for $\widetilde{C}$, by setting $[c_0, \dots, c_{k-1}, c_k]_s$ i-adjacent to $[c_0, \dots, c_{k-1}, c_k']_s$ and $[c_0, \dots, c_{k-1}]_s$ if $c_k' \underset{i}{\sim} c_k$. There is an obvious morphism $\alpha : \overline{C} \to \widetilde{C}$ sending a strict homotopy class to the homotopy class containing it. Using $\psi : (\overline{C}, \bar{c}) \to (C, c)$ for the obvious projection $\psi[c, \dots, d]_s = d$ we have $\varphi\alpha = \psi$

$$(\overline{C}, \bar{c}) \quad \xrightarrow{\alpha} \quad (\widetilde{C}, \tilde{c})$$

$$\psi \searrow \qquad \swarrow \varphi$$

$$(C, c)$$

It suffices to show that ψ is a covering, for then the universality of $\widetilde{C}$ shows α is an isomorphism, so homotopic implies strictly homotopic, and (4.7) does the rest. Thus let R be a rank 2 residue, of type J, in $\overline{C}$; we shall use (R_c) to define a special chamber $z \in R$, and use z to show $\psi|_R$ is an isomorphism. Let $x \in R$ be any chamber; it is a strict homotopy class of galleries in C starting at c, and determines a unique element $w(x) \in W$ (namely r_f where f is the type of such a gallery). Let $w(x) = w'w''$ where w' is the shortest word in the J-residue of W containing w, and $w'' \in W_J$ (this factorization is uniquely determined by J), and let f' and f'' be reduced words with $r_{f'} = w'$, $r_{f''} = w''$. In the class x there is a gallery $\gamma = \gamma'\gamma''$ where γ' has type f', and γ'' has type f''. We define z to be $[\gamma']_s$. If instead of f', f'' we use g', g'', there is a gallery $\delta = \delta'\delta''$ in the class of x, where δ' has type g' and δ'' type g'', and we claim $[\delta']_s = [\gamma']_s$. Indeed δ' and δ'' are strictly homotopic to galleries δ_1' and δ_1'' respectively of types f' and f'', and by (R_c), $\gamma'\gamma'' = \delta_1'\delta_1''$. Therefore γ' is strictly homotopic to δ', and z is well-defined. Moreover, had we started with a chamber $y \underset{i}{\sim} x$, then γ' would be unaffected (only γ'' would change), and so z is uniquely determined by any chamber of R (it is actually $\text{proj}_R\bar{c}$).

If S is the J-residue containing $\psi(R)$ we can now show that $\psi|_R$ is an isomorphism onto S, as required. The existence of z shows that if x is any chamber of R, then $x = [\gamma'\gamma'']_s$, where $[\gamma']_s = z$ and γ'' is a J-gallery of reduced type in S from $\psi(z)$ to $\psi(x)$. Since every chamber $s \in S$ lies at the end of such a gallery γ'' (because S is a rank 2 building), we see that ψ is surjective, and since s determines γ'' uniquely up to strict homotopy (γ'' is in fact unique unless s is opposite $\psi(z)$ in S), ψ is injective.   □

We conclude this section with a beautiful result which is the principal goal of Tits' paper "A Local Approach to Buildings" [1981]. Recall that "spherical type" means W is finite, so a J-residue is of spherical type if W_J is finite.

(4.9) THEOREM. *Let $\widetilde{C}$ be the universal cover of a chamber system C of type M. Then $\widetilde{C}$ is a building if and only if all residues of C of rank 3 and spherical type are covered by buildings.*

PROOF: If $\widetilde{C}$ is a building, then by (3.5) so are its residues, and since these cover the appropriate residues of C, the "only if" part is clear.

To prove the converse we verify that the condition (R_c) of (4.8) is satisfied, so consider a strict homotopy between two galleries of the same reduced type f. This gives a self-homotopy of words, and by (2.17) this decomposes into self-homotopies each of which is either non-essential or lies in a rank 3 residue of spherical type. The former type give an equality of galleries because after a sequence of type $f_1 p(i,j) f_2 \simeq f_1 p(j,i) f_2 \simeq f_1 p(i,j) f_2$ the gallery is left unchanged; and the latter type give an equality of galleries because (R_c) is satisfied in any rank 3 residue of spherical type, by (4.8) and the hypothesis. Therefore the two galleries are equal, and so (R_c) is satisfied. □

(4.10) COROLLARY. *Let C be a chamber system of type M and finite rank ≥ 4, and suppose all residues of C are buildings. If the geometric realization of C is simply-connected in the topological sense, then C is a building.*

PROOF: Since each residue is simply-connected, by (4.3), Exercise 9 shows that C is its own universal cover, and hence a building. □

4. Examples.

In this section we shall look at two examples: a family of chamber systems of type $\widetilde{A}_2$, and an exceptional chamber system of type C_3.

Example 1. $\widetilde{A}_2$ is the rank 3 diagram below for which each $m_{ij} = 3$.

In other words, each of the three types of rank 2 residues is a projective plane (generalised 3-gon), in fact a plane of order 2 in our examples.

First we construct a projective plane of order 2 as follows. Let Frob(21) denote the Frobenius group of order 21; it has a normal Z_7-subgroup with

Z_3 acting non-trivially by conjugation. If P_1 and P_2 are two of its Z_3-subgroups, then using the notation of Example 1 in Chapter 1, the chamber system $(\mathrm{Frob}(21) : B = 1, P_1, P_2)$ is a projective plane (generalized 3-gon) having 21 chambers and 7 panels of each type. This can be verified directly, or indirectly as in Exercise 10, using the fact that $\mathrm{Frob}(21)$ is a subgroup of $SL_3(2)$ acting simple-transitively on the 21 chambers of the building for $SL_3(2)$.

Now let A, B and C be $\mathrm{Frob}(21)$ groups, and take distinct Z_3-subgroups $A_1, A_2 < A$; $B_2, B_3 < B$; $C_3, C_1 < C$. We wish to construct a group G by amalgamating A, B and C so that A_2 becomes identified with B_2, B_3 with C_3, and C_1 with A_1 (see Tits [1986b] regarding amalgams). First notice that if s has order 7, and x has order 3 in $\mathrm{Frob}(21)$, then $xsx^{-1} = s^2$ or s^4. Indeed the two non-identity elements of a Z_3-subgroup of $\mathrm{Frob}(21)$ play different roles: conjugation by one sends each 7-element to its square, conjugation by the other sends it to its fourth power. When we identify A_2 with B_2, etc., we either prescribe a "straight" identification (i.e., the two squaring elements are identified), or a "twisted" one (i.e., each squaring element is identified with the inverse of the other). This distinction yields, up to a reordering of A, B and C, four different amalgamations, which we indicate by the following diagrams:

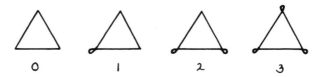

where the corner is straight or twisted in accordance with the identification of the corresponding Z_3-subgroups.

Let $\widetilde{G}_n$ ($n = 0, 1, 2$ or 3), denote the amalgamation of A, B and C in each of the four cases above; it is shown in Tits [1986b] Theorem 1 that this amalgam does not collapse, but contains A, B and C as subgroups (in Cases 0, 1 and 3 this also follows by our construction of quotients of $\widetilde{G}_n$). Now let P_1, P_2 and P_3 denote the Z_3-subgroups of $\widetilde{G}_n$ corresponding to $A_1 = C_1$, $A_2 = B_2$ and $B_3 = C_3$ respectively. Then

$$\widetilde{C}_n = (\widetilde{G}_n : B = 1, P_1, P_2, P_3)$$

is a chamber system of type $\widetilde{A}_2$.

Notice that if G is any group generated by three Z_3-subgroups P_1, P_2, P_3 such that $\langle P_i, P_j \rangle \cong \mathrm{Frob}(21)$, then $C = (G : B = 1, P_1, P_2, P_3)$ is a chamber system of type $\widetilde{A}_2$. Moreover G must be a quotient of $\widetilde{G}_n$ for $n = 0, 1, 2$ or 3, and therefore $\widetilde{C}_n$ covers C. In fact $\widetilde{C}_n$ must be the universal cover (see Exercise 8), and by (4.9), $\widetilde{C}_n$ is a building.

In Cases 0 and 3, Köhler, Meixner and Wester [1984] and [1985] give matrices generating $\widetilde{G}_n$:

Case 0.

$$x = \begin{pmatrix} 1 & 0 & 0 \\ 0 & 0 & 1 \\ 0 & 1 & 1 \end{pmatrix} \quad \tau = \begin{pmatrix} 0 & 1 & t \\ 0 & 1+t & 1 \\ t^{-1}+1+t & t & 1+t \end{pmatrix}$$

$x, \tau \in GL_3(\mathbf{F}_2(t))$.

Case 3.

$$x = \begin{pmatrix} 1 & 0 & -\lambda - 1 \\ 0 & 0 & -1 \\ 0 & 1 & -1 \end{pmatrix} \quad \tau = \begin{pmatrix} 0 & 1 & 0 \\ 0 & 0 & 1 \\ \frac{-\lambda-2}{2} & 0 & 0 \end{pmatrix}$$

where $\lambda^2 + \lambda + 2 = 0$ and $x, \tau \in GL_3(\mathbf{Q}(\sqrt{-7}))$. In either case set $y = x^\tau$ and $z = y^\tau$; then let $P_1 = \langle x \rangle$, $P_2 = \langle y \rangle$ and $P_3 = \langle z \rangle$.

They also show that the group $\widetilde{G}$ generated by x and τ acts transitively on the chambers of the affine building of type $\widetilde{A}_2$ over $\mathbf{F}_2(t)$, or $\mathbf{Q}(\sqrt{-7})$ with the 2-adic valuation (such buildings are dealt with in Chapters 9 and 10), and hence $\widetilde{G}$ is one of the $\widetilde{G}_n$ above. In Case 3 it is natural to try reducing $\widetilde{G}$ modulo a prime p, and this is shown to give:

$$SL_3(\mathbf{F}_p) \text{ if } p = 1, 2 \text{ or } 4(\mathrm{mod} \ 7)$$

$$SU_3(\mathbf{F}_{p^2}) \text{ if } p = 3, 5 \text{ or } 6(\mathrm{mod} \ 7)$$

$$7^2 SL_2(\mathbf{F}_7) \text{ if } p = 7$$

Thus each of these groups acts on a finite chamber system of type $\widetilde{A}_2$, and in each case the universal cover is $\widetilde{C}_3$ (notation above).

In Case 0 one can reduce modulo a prime ideal in $\mathbf{F}_2[t, t^{-1}, (1+t)^{-1}]$. If f is an irreducible polynomial of degree n, reduction mod (f) gives a subgroup of $PGL_3(2^n)$. A result of Köhler-Meixner-Wester [1984], modified by Kantor [1986], is that when $n \geq 10$ this procedure yields more than $2^{n/4}$ different (pairwise non-isomorphic) finite chamber systems of type $\widetilde{A}_2$, each with a group acting simple-transitively on the set of chambers, of course.

In Case 1 a finite example is given in Exercise 12. In Case 2 I do not know of any finite example.

Example 2. This example shows that the rank 3 restriction in Theorem 4.9 is essential. We exhibit a chamber system C of type C_3

$$\underset{1}{\circ}\text{———}\underset{2}{\circ}\text{===}\underset{3}{\circ}$$

which is simply-connected, but not a building. Infinite examples of such objects were discovered years ago by J. Tits. This finite example was first discovered by A. Neumaier, and later independently by M. Aschbacher; it appears as a residue in some higher rank cases (see Exercise 18 for an example).

Take a set of seven elements $1, \ldots, 7$ and call them *points*, and define a *line* to be any subset of three points. Using these points there are exactly 30 ways of choosing seven lines to form a projective plane of order 2, such as the one in Figure 4.2.

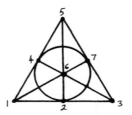

Figure 4.2

This set of planes splits into two orbits under the alternating group A_7: 15 *x-planes* and 15 *y-planes*. Two planes are in the same orbit if and only if they have exactly one line in common (Exercise 14).

To define C let its *chambers* be triples (p, L, X) where p is a point on a line L in an x-plane X; two chambers are 1-, 2- or 3-*adjacent* if they differ in at most one point, line or x-plane respectively. Obviously $\{1,2\}$-residues are projective planes (generalised 3-gons), $\{1,3\}$-residues are generalised 2-gons and, as shown in Exercise 15, $\{2,3\}$-residues are generalised 4-gons, hence the C_3 diagram.

To show C is not a building, count the number of chambers; it is 315. Yet in a C_3 building having 3 chambers per panel, there are 2^9 chambers opposite a given chamber c (9 being the length of the longest word): to see this, count the number of galleries of a given type $i_1 \ldots i_9$ from c to an opposite chamber. Alternatively one could, in the spirit of this chapter, exhibit galleries of types 2132 and 321323 having the same extremities; these words are both reduced, but not homotopic.

To show C is simply-connected, it suffices to show that any closed path in the geometric realization Δ is null-homotopic (see the remark at the beginning of section 2). It is a simple matter (Exercise 17) to reduce to considering paths of the form (p, M, p', M', p) where M and M' are lines on both p and p', and Figure 4.3 shows that such a path (with $p = 1$, $p' = 2$, $M = 123$, $M' = 127$) is null-homotopic. In this picture points, lines and planes are represented as vertices, edges and triangles (it is an exercise to check the planes are x-planes). After making the obvious identifications the illustration shows an octahedron with a slit in one edge, and this is homeomorphic to a disc with the closed path (p, M, p', M', p) as boundary.

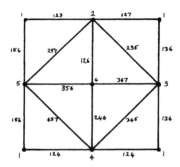

Figure 4.3

Notes. The main results of this chapter are taken from Tits [1981]. In that paper, Tits develops some earlier ideas he had on buildings, which became particularly relevant with the discovery in 1978 of a $\widetilde{G}_2$ geometry (chamber system) for the Lyons sporadic simple group. It was only later that the examples in section 4 were discovered. The paper on amalgams, Tits [1986b], mentioned in Example 1, is similar in spirit to this chapter and is recommended as further reading.

Exercises to Chapter 4

1. Prove Lemma (4.4), and show that if $\varphi : C \to D$ is a covering, and two galleries γ_1, γ_2 in D starting at d are homotopic, then their liftings

$\widetilde{\gamma}_2, \widetilde{\gamma}_2$ to galleries in C starting at $c \in \varphi^{-1}(d)$ must have the same end chamber. Show also that if γ, γ', are homotopic galleries of C, then $\varphi(\gamma), \varphi(\gamma')$ are homotopic, thus verifying injectivity of φ_*.

2. Show that any closed gallery in a building can be reduced to the trivial gallery by a sequence of operations each of which is either a strict elementary homotopy, or an alteration in a rank 1 residue (i.e., $(c, c', c'') \leftrightarrow (c, c'')$ where $c \underset{i}{\sim} c' \underset{i}{\sim} c''$).

3. (Peter M. Johnson). Show that (P_x) is equivalent to (P'_x): The only gallery of reduced type from x to x is the null-gallery. [HINT: With f and f' as in (P_x) use induction on $\min(\ell(f), \ell(f'))$; (P'_x) allows the induction to start].

4. Show that the universal cover of a chamber system of type M is a building if and only if the only closed gallery of reduced type which is null-homotopic is the null-gallery.

In Exercises 5-9, C is any chamber system, not necessarily of type M.

5. Let g be an automorphism of C, and let $\varphi : (\widetilde{C}, \tilde{c}) \rightarrow (C, c)$ be a covering. Show that g lifts to an automorphism $\tilde{g}$ of $\widetilde{C}$ (i.e., $\varphi \circ \tilde{g} = g \circ \varphi$) sending $\tilde{c}$ to $\widetilde{g(c)}$ for some $\widetilde{g(c)} \in \varphi^{-1}(g(c))$ if and only if $(g \circ \varphi)_* \pi(\widetilde{C}, \tilde{c}) = \varphi_* \pi(\widetilde{C}, \widetilde{g(c)})$. [HINT: Use (4.5)].

6. Let $\varphi : \widetilde{C} \rightarrow C$ be a universal covering, and let Π denote the group of automorphisms $\tilde{g}$ of $\widetilde{C}$ which are liftings of the identity (i.e., $\varphi \circ \tilde{g} = \varphi$). Show that Π acts simple-transitively on $\varphi^{-1}(c)$ for any chamber $c \in C$. [HINT: Use Exercise 5 for transitivity].

7. Show that the group Π of Exercise 6 is isomorphic to $\pi(C, c)$. [HINT: Chambers of $\varphi^{-1}(c)$ correspond to homotopy classes of closed galleries based at c].

8. If $\widetilde{C}$ is a universal cover of C, show that any group G of automorphisms of C lifts to a group $\widetilde{G}$ of automorphisms of $\widetilde{C}$ such that $\widetilde{G}/\Pi \cong G$, where Π is the fundamental group of C. Moreover if $\widetilde{R}$ is a residue of $\widetilde{C}$ such that φ maps $\widetilde{R}$ isomorphically onto a residue R of C, then $\mathrm{Stab}_{\widetilde{G}} \widetilde{R} \cong \mathrm{Stab}_G R$.

9. Let C have finite rank $n \geq 3$, and suppose (the geometric realization of) each residue of rank k, for $3 \leq k < n$, is simply-connected in the topological sense. Show that every covering of C is a topological covering. [HINT: A covering restricts to a covering on each residue; use induction on n].

10. Show that the group $SL_3(2)$ has a Frobenius subgroup of order 21 acting simple-transitively on the flags of the projective plane (i.e., $V_1 \subset V_2$ where dim $V_i = i$ in the 3-space on which $SL_3(2)$ acts). Use this to verify the assertion in Example 1 that each $\{i,j\}$-residue is a projective plane.

11. Let P_1, P_2 and P_3 be any three distinct Z_3-subgroups of Frob(21). Show that (Frob(21) $: B = 1, P_1, P_2, P_3$) is a chamber system of type $\tilde{A}_2$, belonging to Case 0 of Example 1.

12. Let $P_1 = \langle x_1 \rangle$, $P_2 = \langle x_2 \rangle$ and $P_3 = \langle x_3 \rangle$ be Z_3-subgroups of the alternating group A_7, where $x_1 = (123)(456)$, $x_2 = (124)(375)$ and $x_3 = (153)(276)$. Show that $(A_7 : B = 1, P_1, P_2, P_3)$ is a chamber system of type $\tilde{A}_2$, belonging to Case 1 of Example 1.

13. What is the universal cover of the rank 3 chamber system derived from S_3 in Exercise 6 of Chapter 1?

14. Verify that, as claimed in Example 2, there are exactly 30 ways of choosing 7 lines to form a projective plane. Show that A_7 has two orbits of size 15 on this set of planes, and that any two distinct planes are in the same orbit if and only if they have exactly one line in common.

15. Define a bipartite graph whose vertices are the *duads* (ab) and *synthemes* $(ab)(cd)(ef)$ of a set $\{a, \dots, f\}$ of six symbols, with incidence being given in the obvious way $[(ab)(cd)(ef)$ incident with $(ab), (cd)$ and $(ef)]$. Show that this is a generalized 4-gon, and corresponds in a natural way to the $\{2,3\}$-residue of Example 2.

16. Treating a generalized 4-gon as a point-line geometry as in Exercise 10 of Chapter 3, show that there is a unique one with parameters $(2,2)$, which is therefore self-dual (i.e., isomorphic to the one obtained by interchanging the roles of points and lines). Conclude, using the preceding exercise, that the symmetric group S_6 admits an outer automorphism interchanging involutions of type (ab) with those of type $(ab)(cd)(ef)$ [it is the only symmetric group admitting an outer automorphism].

17. Let Δ be the geometric realization of the C_3 chamber system of Example 2. Show that if every path in Δ of the form (p, M, p', M', p), where M and M' are lines on points p and p', is null-homotopic then any closed path in Δ can be deformed to a point. [HINT: first deform a closed path to a path consisting of edges of Δ whose vertices are points and lines; any closed path of this form lying in a plane is null-homotopic].

18. In Example 2, let B denote the stabilizer of a chamber c, and let P_1, P_2 and P_3 be the stabilizers of the panels of c (indexed by the diagram, so P_3 stabilizes (p, L), where p is a point on a line L). Show that if $\sigma \in S_7$, then P_2^σ is conjugate to P_2 in A_7, and similarly for P_3, but not for P_1 unless $\sigma \in A_7$. Thus S_7 preserves 2-adjacency and 3-adjacency but not 1-adjacency. Define 1'-adjacency between chambers of C by $c \underset{1'}{\sim} d$ if $\sigma(c) \underset{1}{\sim} \sigma(d)$ for $\sigma \in S_7 - A_7$. Show that this gives a rank 4 chamber system $\widehat{C}$ with diagram

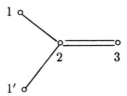

If σ normalizes B, then P_1^σ is the stabilizer of the 1'-panel of c, and $\widehat{C} = (A_7 : B, P_1, P_1^\sigma, P_2, P_3)$. (The $\{1, 2, 1'\}$-residue is the building, of type A_3, for $SL_4(2) \cong A_8$, which admits A_7 as a chamber-transitive automorphism group.)

Chapter 5
BN - PAIRS

This chapter deals with the relation between groups having a Tits system (also called a BN-Pair) and buildings. Parabolic subgroups are defined, and characterised as being those subgroups containing a chamber stabilizer B.

1. Tits Systems and Buildings.

A *Tits System*, or *BN-Pair*, in a group G is a pair of subgroups B, N satisfying:

BN0. $\langle B, N \rangle = G$

BN1. $H = B \cap N \triangleleft N$ and $N/H = W$ is a Coxeter group with distinguished generators $s_1, \ldots, s_n$.

BN2. $BsBwB \subset BwB \cup BswB$ whenever $w \in W$, and $s = s_i$.

BN3. $sBs \neq B$ for $s = s_i$.

Note 1. If $n, n' \in N$ have the same image $w \in W$, then $nB = n'B$, so wB is well defined (cf. BN2).

Note 2. BN0 and BN2 imply that $G = BNB$.

Note 3. Taking inverses in (BN2) and replacing w by w^{-1} gives $BwBsB \subset BwB \cup BwsB$.

(5.1) LEMMA. *(i) If $BwB = Bw'B$ then $w = w'$, and hence G is the disjoint union $\cup BwB$ (called the Bruhat decomposition).*

(ii) If $\ell(sw) > \ell(w)$ then $BsBwB = BswB$.

PROOF: (i) Without loss of generality $\ell(w) \leq \ell(w')$. Let $w = sw_1$ where $\ell(w_1) < \ell(w)$. By assumption $sw_1B \subset Bw'B$. Therefore

$$w_1B \subset sBw'B \subset Bw'B \cup Bsw'B,$$

and so $Bw_1B = Bw'B$ or $Bsw'B$. By induction on $\ell(w)$ (the result obviously being true for $\ell(w) = 0$ - i.e., $w = 1$), we have $w_1 = w'$ or sw'. However $w_1 \neq w'$ because $\ell(w_1) < \ell(w')$, so $w_1 = sw'$, and hence $w = w'$.

(ii) Again using induction on the length of w, we may assume that if $\ell(v) < \ell(w)$, $\ell(sv)$, $\ell(vs')$, then $BsBvB = BsvB$ and $BvBs'B = Bvs'B$ (cf. Note 3). Since $w = vs'$ for some v and s', with $\ell(v) < \ell(w)$, we have

$$BsBwB = BsBvs'B = BsBvBs'B$$
$$= BsvBs'B$$
$$\subset BsvB \cup BswB.$$

Moreover
$$BsBwB \subset BwB \cup BswB.$$

Now $sv \neq w$, otherwise $v = sw$, and since $\ell(sw) > \ell(w)$ by hypothesis, this contradicts $\ell(v) < \ell(w)$; hence $BsvB \neq BwB$ by (i), and the result follows. □

If Δ is a building, a group G of automorphisms of Δ will be called *strongly transitive* if the following two conditions are satisfied:

(i) for each $w \in W$, G is transitive on ordered pairs of chambers (x, y) where $\delta(x, y) = w$.

(ii) there is some apartment Σ whose stabilizer in G is transitive on the chambers of Σ (and hence induces the Coxeter group W on Σ).

In the spherical case, strong transitivity is equivalent to transitivity on the set of all pairs (x, A) where x is a chamber in an apartment A (Exercise 4). In general, however, strong transitivity is a weaker condition; a strongly transitive group need not be transitive on the set of all apartments - an example will appear in Chapter 9 section 2.

The following two theorems explain the connection between strong transitivity and Tits systems. Before stating them we note that if Δ is any chamber system and G a group of automorphisms of Δ acting transitively on the set of chambers, then the chambers of Δ correspond to the left cosets gB where B is the stabilizer of a given chamber (we are taking group action on the left). Thus a double coset BwB consists of those chambers in the same suborbit as wB under the action of B. With this interpretation, axiom BN2 says that s sends a chamber in this suborbit to one either in the same suborbit or in the suborbit containing swB.

(5.2) THEOREM. *Let Δ be a thick building admitting a strongly transitive group G of automorphisms, let Σ be as in the definition of strong transitivity, and let W be the corresponding Coxeter group. Let c be a given chamber of Σ, and let $B = \text{stab}_G c$, $N = \text{stab}_G \Sigma$.*

Then (B,N) is a Tits system, and

$$\delta(c,d) = w \Leftrightarrow d \subset BwB$$

where we are taking d to be a left coset of B.

PROOF: Given $g \in G$ let $w = \delta(c, g(c))$. By strong-transitivity there exists $n \in N$ such that $\delta(c, n(c)) = w$, and $b \in B$ such that $g(c) = bn(c)$, so $g \in bnB \subset BwB$; this proves BN0. Conversely if $g \in BwB$, then $g(c) = bn(c)$ for some $b \in B$, and hence $\delta(c, g(c)) = \delta(c, bn(c)) = \delta(c, n(c)) = w$. Thus we have shown

$$\delta(c,d) = w \Leftrightarrow d = gB \subset BwB.$$

BN1. By (2.2) $B \cap N$ is the kernel of the action of N on Σ, so $B \cap N \triangleleft N$, and by strong transitivity $N/B \cap N \cong W$.

BN2. Let $d = gB \subset BwB$ and suppose $s \in N$ projects to s_i. We need to prove that $\delta(c, s(d)) = w$ or sw. Since $\delta(c, d) = w$, we have $\delta(s(c), s(d)) = w$. Now let c' be the unique chamber nearest to $s(d)$ in the i-residue containing c and $s(c)$ (i.e., $c' = \text{proj}_\pi s(d)$ where π is the panel common to c and $s(c)$). There are three cases:

1) $c' \neq c, s(c)$: therefore $\delta(c, s(d)) = \delta(s(c), s(d)) = w$ (and $\ell(sw) < \ell(w)$)

2) $c' = s(c)$: therefore $\delta(c, s(d)) = s\delta(s(c), s(d)) = sw$ (and $\ell(sw) > \ell(w)$)

3) $c' = c$: therefore $\delta(c, s(d)) = sw$ (and $\ell(sw) < \ell(w)$)

BN3. Using the thickness hypothesis, there is a third chamber d adjacent to both c and $s(c)$; note that $s(d) \neq c$ because $s^2(c) = c$. Now by strong transitivity there exists $b \in B$ sending $s(c)$ to d. Therefore

$$sbs(c) = s(d) \neq c.$$

Therefore $sbs \notin B$.   □

(5.3) THEOREM. *Every Tits system (B,N) in a group G defines a building, the chambers being left cosets of B, with i-adjacency given by*

$$gB \underset{i}{\sim} hB \Leftrightarrow g^{-1}h \in B\langle s_i \rangle B.$$

Moreover $\delta(B, gB) = w \Leftrightarrow gB \subset BwB$ where δ is the distance function on this building, N stabilizes an apartment, and the action of G is strongly transitive.

PROOF: Define $\delta(gB, hB) = w \Leftrightarrow g^{-1}h \in BwB$. We must show there is a gallery of reduced type f from gB to hB if and only if $\delta(gB, hB) = r_f$. Since δ is invariant under group action we restrict attention to $\delta(c, d)$ where $c = B$.

Suppose $\delta(c, d) = w = r_f$ with f reduced. Let $f = jf'$ be reduced, with $j \in I$, and so $w = sw'$ where $s = s_j$ and $w' = r_{f'}$. Without loss of generality $d = wB$, so $s(d) = swB = w'B$. By induction on $\ell(w)$ there is a gallery γ' of type f' from c to $s(d)$, and hence $\gamma = (s(c), \gamma')$ has type $jf' = f$ from $s(c)$ to $s(d)$. Thus $s^{-1}(\gamma)$ is a gallery of type f from c to d.

Conversely suppose there is a gallery $\gamma = (c, c_1, \ldots, d)$ of reduced type $f = jf'$ from c to d; once again set $w = r_f$, $w' = r_{f'}$ and $s = s_j$. Without loss of generality $c_1 = s(c)$, so $s(\gamma) = (s(c), c, \ldots, s(d))$. Thus there is a gallery of type f' from c to $s(d)$, and by induction on $\ell(w)$, $s(d) \subset Bw'B$. Therefore $d \subset s^{-1}Bw'B = sBw'B \subset Bsw'B$ by (5.1). Therefore $\delta(c, d) = sw' = w = r_f$.

Finally let Σ be the set of $n(c)$ for $n \in N$; by (BN1) this is the set of wB as w ranges over W. By definition $\delta(wB, w'B) = w^{-1}w'$, so Σ is isometric to W and is hence an apartment, and N induces W on Σ. Furthermore if $\delta(B, gB) = w$, then $gB = bwB$ for some $b \in B$, so b sends the pair (B, wB) to (B, gB) showing G is strongly transitive. $\qquad\square$

Remarks.

1. The thickness assumption of (5.1) was only used to prove (BN3), and (BN3) was not used in (5.2); so one sees that thickness is equivalent to (BN3).

2. A building Δ with a strongly transitive automorphism group determines a Tits system by (5.2), and by (5.3) this in turn determines a building, which is obviously isomorphic to Δ. On the other hand, given a Tits system in a group G we obtain a building Δ, but G is not uniquely determined by Δ. For example G might be $SL_n(k)$: its centre

acts trivially on Δ and therefore does not appear in Aut Δ (the automorphism group of Δ). Moreover Aut Δ contains more than $PSL_n(k)$;
it is generated by $PGL_n(k)$ together with field automorphisms.

3. Even in a given group there can be more than one Tits system giving
the same building. The subgroup B is uniquely determined because it
has to be a chamber stabilizer, but N can often be replaced by one of
its subgroups (see Exercise 1).

Example. For $G = GL_{n+1}(k)$ take:

$$B = \text{ upper triangular matrices } \begin{bmatrix} * & & * \\ & \ddots & \\ 0 & & * \end{bmatrix}$$

$$N = \text{ matrices with one non-zero entry in each row and}$$

$$\text{column (permutation matrix × diagonal matrix)}$$

$$B \cap N = \text{ diagonal matrices}$$

Here $W \cong S_{n+1}$ with distinguished generators $s_1, \ldots, s_n$, where s_i is the
image of those permutation matrices which are zero off the diagonal except
in positions $(i, i+1)$ and $(i+1, i)$. The building in this case is that of
Example 4 in Chapter 1 (where the chambers are the maximal flags of
projective n-space).

2. Parabolic Subgroups.

For any subset $J \subset I$, recall from Chapter 2 that $W_J = \langle s_j | j \in J \rangle$
(see (2.14)). We now let $P_J = BW_J B$ denote the union of the double
cosets BwB over all $w \in W_J$; by BN2, P_J is a subgroup of G. A *parabolic
subgroup* is a conjugate of one of the P_J; the conjugates of $B = P_\emptyset$ are also
called *Borel subgroups*.

(5.4) THEOREM. *(i) The subgroups containing B are precisely the P_J.*
(ii) $P_J \cap P_K = P_{J \cap K}$, and $\langle P_J, P_K \rangle = P_{J \cup K}$.
(iii) $N_G(P_J) = P_J$, and P_J is the stabilizer of the J-residue containing c.
*(iv) There is a bijection of double coset spaces $W_J \backslash W / W_K \to P_J \backslash G / P_K$
defined by $W_J w W_K \mapsto P_J w P_K$.*

PROOF: (i) Let P be a subgroup containing B, and let $J = \{j \in I | Bs_j B \subset
P\}$; we claim $P_J = P$. Since P_J is generated by the $Bs_j B$ for $j \in J$
(immediate from (5.1)(ii)), we have $P_J \subset P$. Conversely suppose $BwB \subset P$
and let $w = sw'$ where $\ell(w') < \ell(w)$. Since $BwB = BsBw'B$ it suffices, by

induction on $\ell(w)$, to show $BsB \subset P$; to do this let d be a third chamber in the rank 1 residue containing c and $s(c)$ (this uses (BN3)) - see Figure 5.1.

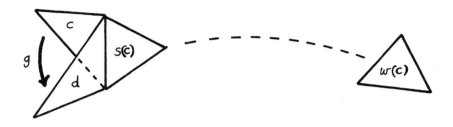

Figure 5.1

The stabilizer of $w(c)$, namely wBw^{-1}, is transitive on chambers at some given distance from $w(c)$, and since $\delta(c, w(c)) = w = \delta(d, w(c))$, there exists $g \in wBw^{-1} \subset P$ sending c to d. Since $\delta(c, d) = s$ we have $g \in BsB$, hence $BsB \subset P$, completing the proof.

(ii) The fact that $P_J \cap P_K = P_{J \cap K}$ follows immediately from $W_J \cap W_K = W_{J \cap K}$ (Exercise 3 of Chapter 2). To see that $\langle P_J, P_K \rangle = P_{J \cup K}$ we note that by (i), $\langle P_J, P_K \rangle = P_L$ for some L containing J and K; but $P_J, P_K \subset P_{J \cup K}$, so $L \subset J \cup K$, and hence $L = J \cup K$.

(iii) Let R be the J-residue containing c; its chambers are all x for which $\delta(c, x) \in W_J$, and so for $g \in G$, we have

$$g(R) = R \Leftrightarrow \delta(c, g(c)) \in W_J \Leftrightarrow g \in P_J.$$

To show these P_J are self-normalizing, note that by (i) $N_G(P_J) = P_K$ for some $K \supset J$. If $i \in K$ then $s_i B s_i = s_i B s_i^{-1} \subset P_J$. Moreover by BN2 and BN3, $s_i B s_i \cap B s_i B \neq \emptyset$ and therefore P_J must contain the double coset $B s_i B$, showing $i \in J$ and so $J = K$.

(iv) Using (5.1)(ii) it is a straightforward exercise to see that $P_J w P_K = B W_J w W_K B$, and hence the map $W_J w W_K \mapsto P_J w P_K$ is well-defined. Moreover by (5.1)(i) it is both injective and surjective.   □

Example. In the example of $GL_{n+1}(k)$ above, let $J = \{t, t+1, \ldots, t+m-1\} \subset \{1, \ldots, n\} = I$. Then P_J has $GL_{m+1}(k)$ as a quotient group, and consists of all matrices, as shown below, which are zero below the diagonal

except in the $m \times m$ block whose top left corner occupies the diagonal
position (t, t).

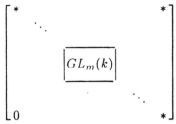

Notes. The axioms for a BN-Pair were given by Tits [1962]; for the genesis
of these ideas see Tits [1974] page IX, where work of Curtis [1964] is also
mentioned. The subgroups B and N appear in a natural way in the theory
of linear algebraic groups: B is a maximal, connected, solvable subgroup
(called a Borel subgroup, after A. Borel), and N is the normalizer of a
torus.

Exercises to Chapter 5

1. In the $GL_{n+1}(k)$ example of this chapter, let N_0 be the group of permutation matrices. Show that (B, N_0) is also a BN-pair for G determining
   the same building as (B, N); notice that $B \cap N_0 = 1$, and $N_0 \cong W$.

2. Set $H_1 = \bigcap_{n \in N} nBn^{-1}$. A BN-Pair (B, N) is called *saturated* if $H_1 = B \cap N$. If Σ is the apartment stabilized by N, show that $\text{Stab}_G \Sigma = N$
   if and only if (B, N) is saturated. In general show that (B, NH_1) is
   saturated, and determines the same building as (B, N).

3. Let K be the kernel of the action of G on the building determined by
   a Tits system (B, N). Show that K is the largest normal subgroup of
   G contained in B.

4. Show that for buildings of spherical type, strong transitivity is equivalent to transitivity on pairs (x, A) where x is a chamber in an apartment A.

5. If P_J is conjugate to P_K show that $J = K$.

6. If $\nu : N \longrightarrow W$ is the natural projection, let $N_J = \nu^{-1}(W_J)$. Show that
   $N_J = N \cap P_J$ and that (B, N_J) is a BN-pair for P_J.

7. Let K be a normal subgroup of G. If $BK = P_J$ then show that for
   $i \in I - J$ and $j \in J$ one has $m_{ij} = 2$ (i.e., s_i and s_j commute). In

particular if the diagram is connected then either $K < B$ or $BK = G$. [HINT: Show $Bs_j \cap K \neq \emptyset$, hence $s_i^{-1}Bs_j s_i \cap P_J \neq \emptyset$, and therefore $Bs_j s_i \cap s_i BwB \neq \emptyset$ for some $w \in W_J$; apply (5.1)].

8. Suppose G has a Tits system (B, N) with a connected diagram, and suppose B is solvable and G perfect. Show that any normal subgroup of G lies in B; moreover if G acts faithfully on the building determined by (B, N) then G is simple.

9. Let Δ be the building for $GL_3(k)$, i.e. vertices are 1- and 2-spaces of a 3-dimensional vector space V over k (and edges are given by containment); here $W \cong D_6$. Then let Δ' denote the barycentric subdivision of Δ (i.e., obtained by interposing an additional vertex in the middle of each edge of Δ); Δ' is a non-thick building with $W \cong D_{12}$. Let σ be an isomorphism switching V with its dual; obviously σ acts on Δ switching 1-spaces with 2-spaces. Let G be the group generated by σ and $GL_3(k)$. Show that G acts strongly transitively on Δ' and use this to verify that G has a "weak Tits system", i.e., satisfying BN0, BN1, BN2, but not BN3.

Chapter 6
BUILDINGS OF SPHERICAL TYPE AND ROOT GROUPS

A building of *spherical type* is one for which W is finite (so each apartment is a triangulation of a sphere - see Chapter 2 section 4). A powerful theorem in Chapter 4 of Tits [1974], repeated here without proof as (6.6), shows that a spherical building admits non-trivial automorphisms when the rank is at least three. Moreover if each connected component of the diagram has rank at least 3, this implies (6.7) that a thick spherical building necessarily admits "root groups", and these generate a group with a BN-pair. All buildings in this chapter will be thick, and also spherical, except in section 4 when we discuss a generalization of "root groups" to non-spherical buildings.

1. Some Basic Lemmas.

An important fact about finite Coxeter complexes W is that every chamber has a unique opposite, and W is the convex hull of any two opposite chambers (see Theorem (2.15)). In a spherical building two chambers are called *opposite* if they are opposite in some apartment containing them, in which case this apartment is unique as it is the convex hull of x and y (cf. Exercise 5 in Chapter 3). Notice that if $d = \mathrm{diam}(W)$, then x and y are opposite if and only if $d(x, y) = d$.

One of the first things we want to do is to extend the idea of opposites to all simplexes of a spherical building; we first deal with panels.

(6.1) LEMMA. *Let π be a panel (of type i) on chambers x and x' in an apartment A. If y and y' denote the chambers of A opposite x and x' respectively, then y and y' are adjacent. Moreover if π' is the panel (of* *type i') common to y and y', then π and π' determine the same wall of A, and $r_{i'} = w_o^{-1} r_i w_o$ where w_o is the longest word of W.*

PROOF: Since x' is opposite y' and adjacent to x, we have $d(x, y') = d - 1$. Moreover by (2.15) (iv) y' lies on a minimal gallery from x to y, so

$$d(y', y) = d(x, y) - d(x, y') = 1$$

showing y and y' are adjacent. Now if α is the root of A containing x but not x', then $y \in -\alpha$ and $y' \in \alpha$ so the wall $\partial \alpha$ contains both π' and π.

For the final statement, treat A as the Coxeter complex W, in which case $x' = x r_i$, $y = x w_o$, $y' = x' w_o$ and $y' = y r_{i'}$. Thus $x w_o r_{i'} = x r_i w_o$, so $r_{i'} = w_o^{-1} r_i w_o$. □

For an apartment A, let $\mathrm{op}_A : A \to A$ denote the map sending each chamber of A to its opposite; we call it the *opposition involution*. Although op_A is not necessarily an automorphism, Lemma (6.1) shows that it sends i-adjacency to i'-adjacency where $r_i w_o = w_o r_{i'}$. It is an automorphism if and only if $i' = i$ for all $i \in I$, in which case $\mathrm{op}_W = w_o \in W$. Notice that the opposition involution induces a symmetry of the diagram, and so $\mathrm{op}_W = w_o$ whenever the diagram exhibits no non-trivial symmetry (e.g., C_n for $n \geq 3$). For types A_n and E_6 it reverses the diagram (Exercise 1), and for D_n it induces a non-trivial symmetry precisely when n is odd (Exercise 2). Finally, as mentioned in Chapter 2 section 4, a finite Coxeter group W preserves a dot product on $\mathbf{R}^n$, and the Coxeter complex can be taken as a triangulation of the $(n-1)$-sphere S^{n-1}; op_W is then simply the antipodal map, sending v to $-v$ for all $v \in \mathbf{R}^n$.

We now define two simplexes of W to be *opposite* if they are interchanged by op_W. More generally, two simplexes of a spherical building are *opposite* if they are opposite in some apartment containing them (hence in every such apartment, by Exercise 7 of Chapter 3).

(6.2) LEMMA. *Given opposite panels π and π' in a spherical building, and chambers $x \in St(\pi)$ and $y \in St(\pi')$ one has $d(x, y) = d$ unless $x = \mathrm{proj}_\pi y$ in which case $d(x, y) = d - 1$. In particular $\mathrm{proj}_\pi | St(\pi')$ is inverse to $\mathrm{proj}_{\pi'} | St(\pi)$. (Recall that $\mathrm{proj}_\pi x$ is the unique chamber of $St(\pi)$ nearest x).*

PROOF: It suffices to show that x is opposite some (hence all but one) chamber on π', but of course if A is any apartment containing x and π', then x is opposite $\mathrm{op}_A(x)$ which has $\pi' = \mathrm{op}_A(\pi)$ as a panel. □

Recall that a *root* of a building is a root in an apartment of that building (i.e., a half-apartment).

(6.3) LEMMA. *Let α be a root in a spherical building, and x a chamber having a panel π in $\partial\alpha$. Then there is a unique root containing x and $\partial\alpha$, and if $x \notin \alpha$ there is a unique apartment containing x and α.*

PROOF: By (6.1) $\partial\alpha$ contains a panel π' opposite π. Let $y' = \mathrm{proj}_{\pi'}x$, and let y denote the chamber of α on π', so $x' = \mathrm{proj}_{\pi}y$ is the chamber of α on π - see Figure 6.1.

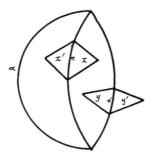

Figure 6.1

By (6.2) $d(x, y') = d - 1 = d(x', y)$, so by Exercise 5 the convex hull of x and y' is a root we call β, and similarly α is the convex hull of x' and y. If $x \notin \alpha$, then by (6.2) x is opposite y, and we let A be the unique apartment containing x and y. Since A contains x' it contains α, completing the proof.

□

Before stating our next proposition we introduce the notation $E_1(c)$ to mean the set of chambers adjacent to c.

(6.4) PROPOSITION. *Let c and b be opposite chambers of a spherical building (assumed to be thick of course), and suppose φ is an automorphism fixing b and all chambers of $E_1(c)$. Then φ is the identity.*

PROOF: By connectivity (and induction along a gallery from b) it suffices to show that if $b' \sim b$, then φ fixes b' and all chambers of $E_1(c')$ for some c' opposite b'. Let π be the common panel of b and b', and let σ be the panel of c opposite π. By (6.2) $b' = \mathrm{proj}_{\pi}x$ for some $x \in St(\sigma)$, hence φ fixes b';

and since b' may be chosen arbitrarily φ fixes all chambers of $E_1(b)$. If b' is opposite c we are done, so suppose not. Using the thickness assumption there exists a chamber c' of $St(\sigma)$ opposite both b and b' (namely any $c' \in St(\sigma)$ with $c' \neq \text{proj}_\sigma b$ or $\text{proj}_\sigma b'$). Since φ fixes c' and $E_1(b)$, the argument above shows it fixes $E_1(c')$, concluding the proof.   □

Remark. Without the thickness assumption (always valid in this Chapter) the above Proposition is false; see Exercise 18.

2. Root Groups and the Moufang Property.

For any root α (i.e., half-apartment) in a spherical building, let

$$U_\alpha = \{g \in \text{Aut } \Delta \mid g \text{ fixes every chamber having a panel in } \alpha - \partial\alpha\}.$$

This will be called a *root group* if the diagram has no isolated nodes. In fact given this condition on the diagram there is a chamber $c \in \alpha$ such that no panel of c is in $\partial\alpha$ (see Exercise 7), or equivalently such that every chamber of $E_1(c)$ has a panel in $\alpha - \partial\alpha$. Since any apartment A containing α contains a chamber opposite c, it is immediate from (6.4) that only the identity of U_α fixes A. If for each root α, U_α is transitive, and hence simple-transitive, on the set of apartments containing α, we call the building *Moufang*. In fact it suffices to assume this condition for the roots α in a given apartment Σ. Indeed if g sends α to β, then $gU_\alpha g^{-1} = U_\beta$, and it is not difficult to show that for those α in Σ the U_α generate a group having a BN-Pair (see 6.16); in the spherical case such a group is transitive on the set of apartments, hence on the set of roots, so each U_β acts in the required manner. Notice that by (6.3) the set of apartments containing α corresponds bijectively to the set of chambers $x \notin \alpha$ on some given panel π of $\partial\alpha$; if the chambers of $St(\pi)$ correspond to the points of a projective line, as they do in many cases, then U_α is the translation group of this line, isomorphic to the additive group of the field (cf. Example 1 below).

(6.5) LEMMA. *If α contains a chamber c having no panel in $\partial\alpha$, and if U_α is transitive on apartments containing α (e.g. as in the Moufang case), then*

$$U_\alpha = \{g \in \text{Aut } \Delta \mid g \text{ fixes } \alpha \text{ and every chamber of } E_1(c)\}.$$

PROOF: Exercise.   □

Example 1. Let Δ be the building of Example 4 in Chapter 1, where the chambers are the maximal flags $V_1 \subset \ldots \subset V_n$ of an $(n+1)$-dimensional vector space V over k. As explained in Chapter 1, a basis $e_1, \ldots, e_{n+1}$ of V determines an apartment A of Δ, whose chambers are all maximal flags $\langle e_{\sigma(1)} \rangle \subset \ldots \subset \langle e_{\sigma(1)}, \ldots, e_{\sigma(n)} \rangle$ as σ ranges over S_{n+1}. The reflection r switching e_i with e_j determines two opposite roots of A, which we call α and $-\alpha$. Writing matrices with respect to the basis $e_1, \ldots e_{n+1}$, the root groups U_α and $U_{-\alpha}$ are (after possibly interchanging α and $-\alpha$) the groups of matrices having 1 in each diagonal position and 0 in every other position except the (i,j) position for U_α, and the (j,i) position for $U_{-\alpha}$ (see Exercise 11). Notice that U_α and $U_{-\alpha}$ are isomorphic to the additive group of the field k. $\qquad\square$

Extending the $E_1(c)$ notation, we let $E_2(c)$ denote the set of chambers lying in one of the rank 2 residues containing c - i.e. having a face of codimension 2 in common with c. If Δ has rank 2, then of course $E_2(c) = \Delta$. The following very strong theorem is (4.16) of Tits [1974], and we shall not prove it here.

(6.6) THEOREM. *Let A and A' be apartments containing chambers c and c', in spherical buildings Δ and Δ' respectively. Then any isomorphism from $E_2(c) \cup A$ to $E_2(c') \cup A'$ extends to an isomorphism from Δ to Δ'.*

$\qquad\square$

(6.7) COROLLARY. *If Δ is a spherical building such that each connected component of the diagram has rank ≥ 3, then Δ is Moufang.*

PROOF: The condition on the diagram ensures that any root α contains a chamber c none of whose faces of codimension 1 or 2 lies in $\partial\alpha$ (see Exercise 8), in which case $E_2(c) \cap A \subset \alpha$ for any apartment A containing α. Therefore if A and A' are apartments containing α, then the isomorphism from A to A' fixing α must also fix $E_2(c) \cap A$, and hence can be extended by the identity to an automorphism from $E_2(c) \cup A$ to $E_2(c') \cup A'$. By Theorem 6.6 this extends to an automorphism g of Δ fixing α and all chambers of $E_2(c)$, and sending A to A'. It remains to show that g fixes all chambers having a panel π in $\alpha - \partial\alpha$; this is Exercise 9. $\qquad\square$

Not all rank 2 spherical buildings are Moufang; for example many non-Moufang projective planes are known, and the generalized quadrangle constructed in Exercise 20 of Chapter 3 is not Moufang if k is a field with at least 4 elements. Moreover Exercise 21 of Chapter 3 constructed "free" generalized m-gons and these have zero probability of being Moufang. However

the following theorem eliminates non-Moufang m-gons from consideration
as residues in higher rank cases.

(6.8) THEOREM. *If Δ is Moufang, so is every residue whose residual subdiagram does not contain isolated nodes (so that the term Moufang applies).*

PROOF: Let R be such a residue; α_o a root of R, and $\pi \in \partial\alpha_o$ a panel of
the chamber $c \in \alpha_o$; we must show U_{α_o} is transitive on $St(\pi) - \{c\}$. By
(3.6), α_o lies in an apartment A of Δ (if R has type J, α_o is isometric to a
subset of W_J, hence also of W), and if α denotes the root of A such that
$c \in \alpha$ and $\pi \in \partial\alpha$, then $\alpha_o \subset \alpha$, and $U_\alpha \subset U_{\alpha_o}$. Now since Δ is Moufang,
U_α is transitive on $St(\pi) - \{c\}$ (and in fact $U_\alpha = U_{\alpha_o}$), completing the
proof.   □

The following remarkable theorem was proved by Tits [1976/79] and
Weiss [1979] (see Appendix 1 for more details).

(6.9) THEOREM. *(Tits-Weiss): Moufang generalized m-gons can exist only
for $m = 3, 4, 6$ and 8.*

PROOF: Given in Appendix 1.   □

Remark. There do indeed exist Moufang m-gons for $m = 3, 4, 6$ and 8
(see Appendix 2 for more details).

(6.10) COROLLARY. *There is no (thick) building whose diagram has an
H_3 (i.e. ○——○ $\overset{5}{}$○) subdiagram.*

PROOF: By (6.7) an H_3 residue is Moufang, and by (6.8) it contains Moufang 5-gons, which do not exist.   □

3. Commutator Relations.

In this section we consider the commutator $[U_\alpha, U_\beta]$ of two root groups,
but first we prove a lemma in the rank 2 case. Let Σ be an apartment of a
Moufang m-gon (i.e., Σ is a $2m$-gon), let $\wedge$ be a gallery of Σ having at least
three chambers, and let c be an interior chamber of $\wedge$. If $\wedge$ is contained in
a root, let $U_1, \ldots, U_k$ be the root groups in a natural cyclic order for those
roots of Σ containing $\wedge$.

(6.11) LEMMA. *The group X fixing $\wedge$ and all chambers of $E_1(c)$ is the
product $U_1 \ldots U_k$, unless $\wedge$ lies in no root in which case X is the identity.*

PROOF: If $\wedge$ lies in no root then Σ is the only apartment containing it;
in this case X fixes $E_1(c)$ and a chamber opposite c, and is therefore the

identity by (6.4). Now suppose $\wedge$ lies in a root α; then U_α fixes $\wedge$ and $E_1(c)$, so $U_1 \ldots U_k \subset X$. Conversely let $g \in X$. If v is an end vertex of $\wedge$, and $w \notin \wedge$ the next vertex in Σ (see Figure 6.2), then there exists $u \in U_1$ (or U_k, but without loss of generality we take it to be U_1) such that $u(w) = g(w)$.

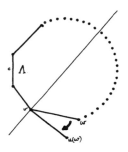

Figure 6.2

Letting $\wedge'$ denote $\wedge$ plus the edge vw, we see that $u^{-1}g$ fixes $\wedge'$ and all chambers of $E_1(c)$; by a simple induction $u^{-1}g \in U_2 \ldots U_k$ and hence $X \subset U_1 \ldots U_k$.   □

Now let Σ be an apartment in a spherical building, and let Φ denote the set of roots of Σ. For roots $\alpha, \beta \in \Phi$ such that $\alpha \neq \pm\beta$, we set

$$[\alpha, \beta] = \{\gamma \in \Phi \mid \alpha \cap \beta \subset \gamma\}.$$

Regarding Σ as a sphere (see the remarks on sphericity in Chapter 2 section 4), its roots are hemispheres and the condition $\alpha \neq \pm\beta$ implies that the walls $\partial\alpha$ and $\partial\beta$ intersect (transversely), and hence $\partial\alpha \cap \partial\beta$ has codimension 2. If σ is any codimension 2 simplex in $\partial\alpha \cap \partial\beta$, then its opposite σ' also lies in $\partial\alpha \cap \partial\beta$, and if $\gamma \in [\alpha, \beta]$ then σ and σ' lie in the wall $\partial\gamma$ (see Exercise 4). Thus $\gamma \cap St(\sigma)$ is a root in the rank 2 apartment $\Sigma \cap St(\sigma)$. Moreover γ is the unique root of Σ containing $\gamma \cap St(\sigma)$ and with $\sigma \in \partial\gamma$, so there is no loss in considering $[\alpha, \beta]$ in the rank 2 residue $St(\sigma)$ (more precisely $[\alpha, \beta] \cap St(\sigma) = [\alpha \cap St(\sigma), \beta \cap St(\sigma)]$). See Figure 6.3 for an illustration in the rank 3 case:

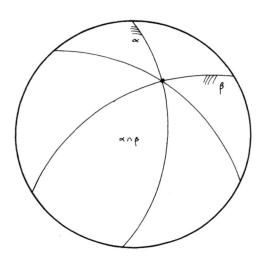

Figure 6.3

If the rank of Σ is greater than 3, then $\partial\alpha \cap \partial\beta$ is connected and there are several choices for σ and σ'.

We next observe that since the elements of U_γ are uniquely determined by their action on the chambers of $St(\pi)$ for any panel $\pi \in \partial\gamma$, we may take $\pi \in St(\sigma)$, in which case it is clear that U_γ is identical to the root group $U_{\gamma \cap St(\sigma)}$ defined on the rank 2 residue $St(\sigma)$. Furthermore if $g \in \langle U_\gamma \mid \gamma \in [\alpha, \beta]\rangle$ is the identity on $St(\sigma)$, then it is the identity on Δ (because each U_γ fixes the simplex of Σ opposite σ, so by Exercise 6, g fixes Σ, and by (6.4) g is the identity). In particular when we consider the commutator $[U_\alpha, U_\beta]$ we may restrict our attention to a single rank 2 residue.

Finally we set $(\alpha, \beta) = [\alpha, \beta] - \{\alpha, \beta\}$, and for any set Ψ of roots we write $U_\Psi = \langle U_\alpha \mid \alpha \in \Psi\rangle$.

(6.12) THEOREM. *In a Moufang building of spherical type, one has:*

(i) for roots $\alpha, \beta \in \Phi$ with $\alpha \neq \pm\beta$

$$[U_\alpha, U_\beta] \leq U_{(\alpha,\beta)}$$

(ii) Let $\{\beta_1, \dots, \beta_k\} = [\beta_1, \beta_k]$ in the natural cyclic order. Then the commutator relation above implies that

$$U_{[\beta_1, \beta_k]} = U_{\beta_1} \dots U_{\beta_k}$$

In particular $U_{\beta_1} \dots U_{\beta_k} = U_{\beta_k} \dots U_{\beta_1}$.

PROOF: (i) As observed above it suffices to work in the rank 2 case (i.e. in an apartment of $St(\sigma)$ for some σ in $\partial\alpha \cap \partial\beta$). In this rank 2 apartment Σ, let x and y be the end vertices of $\alpha \cap \beta$ with x in the interior of α - see Figure 6.4.

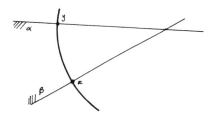

Figure 6.4

If e is a chamber on x, then U_α fixes e; hence $[U_\alpha, U_\beta]$ fixes e, and similarly any chamber on y. Now let $\wedge$ denote $\alpha \cap \beta$ plus the other chamber in Σ on x, and that on y. Certainly $\wedge$ contains at least one interior chamber c, and $[U_\alpha, U_\beta]$ fixes all of $E_1(c)$ and $\wedge$. Since (α, β) is the set of roots containing $\wedge$, (6.11) implies $[U_\alpha, U_\beta] \leq U_{(\alpha, \beta)}$.

   (ii) By induction on k, it suffices to show that if $u_i \in U_{\beta_i}$ for $i = 1, \dots, k$, then $u_k u_1 \dots u_{k-1} \in U_{\beta_1} \dots U_{\beta_k}$. By part (i) $u_k u_1 \dots u_{k-1} = u_1 u_k v u_2 \dots u_{k-1}$ where $v \in U_{\beta_2} \dots U_{\beta_{k-1}}$, and the induction hypothesis shows $v u_2 \dots u_{k-1} = v_2 \dots v_{k-1}$ where $v_i \in U_{\beta_i}$. Repeating this procedure on $u_k v_2 \dots v_{k-1}$, an obvious induction completes the proof.   □

   Our next proposition shows that if α and β are roots containing a chamber c with a panel in $\partial\alpha$ and a panel in $\partial\beta$, then $[U_\alpha, U_\beta]$ is non-trivial, unless the walls $\partial\alpha$ and $\partial\beta$ are perpendicular. More precisely consider a rank 2 apartment Σ having $2m$ chambers, with $m \geq 3$, and let c be a chamber of Σ. If $\beta_1, \dots, \beta_m$ denote the roots of Σ containing c, in one of two natural cyclic orders, then for $x \in U_{\beta_1}$ and $y \in U_{\beta_m}$, (6.12) yields

$$[x, y] = z_2 \dots z_{m-1}$$

where $z_t \in U_{\beta_t}$. We shall write $[x, y]_t$ for z_t.

(6.13) PROPOSITION. *Let $m \geq 3$ and let $x \in U_{\beta_1} - \{1\}$. Then with the notation above, $y \mapsto [x, y]_2$ is an isomorphism from U_{β_m} to U_{β_2}. Furthermore if $m = 3$ then the root groups are abelian.*

PROOF: Let d, e be the chambers of $\beta_2 - \beta_1$, $\beta_3 - \beta_2$ respectively, and let π be the panel common to d and e - see Figure 6.5; we also define $d' = x^{-1}(d)$ and $\pi' = x^{-1}(\pi)$.

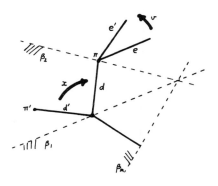

Figure 6.5

We claim first that U_{β_m} acts simple-transitively on $St(\pi') - \{d'\}$. Indeed if $g \in U_{\beta_{m+1}}$ sends d' to c, then $gU_{\beta_m}g^{-1} = U_{\beta_m}$ by (6.12), and hence U_{β_m} acts the same way on $St(\pi') - \{d'\}$ as it does on $St(g(\pi')) - \{c\}$, namely simple-transitively.

Now let $v \in U_{\beta_2}$ be any element, and set $e' = v(e)$. We shall find $y \in U_{\beta_m}$ such that $[x, y]_2$ sends e to e', hence $[x, y]_2 = v$. First notice that for $t > 2$, $e \in \beta_t$ and so U_{β_t} fixes e, and we have $[x, y]_2(e) = [x, y](e) = xyx^{-1}(e)$. Now since U_{β_m} is simple-transitive on $St(\pi') - \{d'\}$, $xU_{\beta_m}x^{-1}$ is simple-transitive on $St(\pi) - \{d\}$, so there is a unique $y \in U_{\beta_m}$ with $[x, y]_2 = v$, and $y \mapsto [x, y]_2$ is an isomorphism from U_{β_m} to U_{β_2}.

Finally let $m = 3$, and let $u, v \in U_{\beta_2}$. Then $v = [x, y]$ for suitable $x \in U_{\beta_1}$, $y \in U_{\beta_3}$, and since U_{β_2} commutes with U_{β_1} and U_{β_3}, we have $[u, v] = [u, [x, y]] = 1$.   □

Remark 1. Let α be any root in a Moufang building (of spherical type), and let π be a panel of type i in the wall $\partial \alpha$. If the i-node of the diagram lies on a single bond and so $m_{ij} = 3$ for some $j \in I$, then π has a face σ of type $\{i, j\}$, and by (6.13) the root group U_α is abelian. This is the case whenever each connected component of the diagram has rank at least 3 and is not of type C_n (see Appendix 5 for diagrams); for example F_4 buildings have two conjugacy classes of root groups both of which are abelian (see Exercise 13). In the C_n case there are two types of roots in one of which all panels in the boundary wall $\partial\alpha$ have type n, where n is the end node on the double bond of the diagram; in this case U_α is not necessarily abelian.

Remark 2. For the case $m = 3$, (6.12) and (6.13) give complete information on the commutator $[U_\alpha, U_\beta]$ when $\beta \neq -\alpha$. For $m = 4, 6$ or 8 see Tits [1976a] and [1983] for further details.

4. Moufang Buildings - the general case.

In this section we consider subgroups generated by root groups, and define an analogue of the Moufang condition for thick buildings which are not necessarily of spherical type. Let Φ be the set of roots in a given apartment Σ (not necessarily of spherical type!). Following Tits [1987] we call a pair of roots $\alpha, \beta \in \Phi$ *prenilpotent* if both $\alpha \cap \beta$ and $(-\alpha) \cap (-\beta)$ are non-empty sets of chambers. In the spherical case this is equivalent to saying $\alpha \neq -\beta$, and in the general case it means that either the walls $\partial\alpha$ and $\partial\beta$ intersect and are distinct, or else $\alpha \subset \beta$ or $\beta \subset \alpha$ (Exercise 14). Given such a pair α, β we write

$$[\alpha, \beta] = \{\gamma \in \Phi \mid \alpha \cap \beta \subset \gamma \text{ and } (-\alpha) \cap (-\beta) \subset -\gamma\}.$$

This set $[\alpha, \beta]$ is finite (Exercise 15), and in Figure 6.6 we illustrate the generic rank 3 case with $\beta \subset \alpha$, in which roots are half-spaces of the hyperbolic plane (a rank 3 Coxeter complex, containing no o__∞__o subdiagram, and which is not spherical or affine (see Chapter 9) is a triangulation of the hyperbolic plane).

Figure 6.6

In the spherical case the condition $\alpha \cap \beta \subset \gamma$ implies $(-\alpha) \cap (-\beta) \subset -\gamma$

(see Figure 6.3), so $[\alpha, \beta]$ agrees with the definition given earlier in this case. As before we set $(\alpha, \beta) = [\alpha, \beta] - \{\alpha, \beta\}$.

We now define Δ to be *Moufang* if there is a set of groups $(U_\alpha)_{\alpha \in \Phi}$ satisfying the following conditions, in which case the U_α are called *root groups*.

(M1) If π is a panel of $\partial\alpha$, and c is the chamber of $St(\pi)$ in α, then U_α fixes all the chambers of α and acts simple-transitively on $St(\pi) - \{c\}$.

(M2) If $\{\alpha, \beta\}$ is a prenilpotent pair of distinct roots, then $[U_\alpha, U_\beta] \leq U_{(\alpha, \beta)}$.

(M3) For each $u \in U_\alpha - \{1\}$ there exists $m(u) \in U_{-\alpha}\, u\, U_{-\alpha}$ stabilizing Σ (i.e. interchanging α with $-\alpha$).

(M4) If $n = m(u)$ then for any root β, $nU_\beta\, n^{-1} = U_{n\beta}$.

This definition of a Moufang building is given by Tits [1987] p.563. As shown below, it agrees with the earlier definition of a Moufang building when the diagram is of spherical type having no isolated nodes. In general however α will not uniquely determine U_α - there can be many choices for systems $(U_\alpha)_{\alpha \in \Phi}$ satisfying (M1) - (M4); see Chapter 9 section 2. Notice however that given $u \in U_\alpha - \{1\}$ there is a unique $m(u) = vuv'$ where $v, v' \in U_{-\alpha}$ as in (M3); this follows from the simple-transitivity of U_α and $U_{-\alpha}$.

Examples. Not all buildings admit a system $(U_\alpha)_{\alpha \in \Phi}$, even if they admit a group with a BN-Pair; a good example is the affine building for $SL_n(Q_p)$ where $n \geq 3$ and Q_p is the p-adic numbers (described in Chapter 9 section 2). Examples of Moufang buildings which are not of spherical type come from Kac-Moody groups (see Tits [1987]); in the affine case they arise from algebraic groups over function fields, such as $SL_n(k(t))$ (again see Chapter 9 section 2).

(6.14) PROPOSITION. *A root group U_α fixes every chamber having a panel in $\alpha - \partial\alpha$. In particular in the spherical case the U_α are root groups in the earlier sense, and moreover satisfy (M1) - (M4).*

PROOF: Let $x \notin \alpha$ be a chamber having a panel π in $\alpha - \partial\alpha$, and let $u \in U_\alpha$ be any element. The wall of Σ containing π determines two opposite roots; let β be the one whose opposite $-\beta$ has non-empty intersection with $-\alpha$, so $\{\alpha, \beta\}$ is a prenilpotent pair - see Figure 6.7.

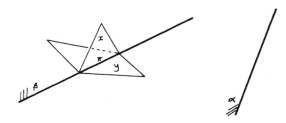

Figure 6.7

If y is the unique chamber of $St(\pi) \cap (-\beta)$, then $y = v(x)$ for some $v \in U_\beta$. For $\gamma \in [\alpha, \beta]$, with $\gamma \neq \beta$, we have $y \in \gamma$, so U_γ fixes y, and by (M2)

$$v(x) = y = [v, u](y) = vuv^{-1}(y) = vu(x).$$

Therefore $u(x) = x$, proving the first statement.

For the last statement, let the U_α be root groups in a Moufang building of spherical type. Then (M1) follows from the definition, as explained at the beginning of section 2, (M2) is (6.12) (i), and (M4) follows from the fact that α uniquely determines U_α. To prove (M3), let π be a panel of $\partial\alpha$, and let c, c' be the chambers of $St(\pi)$ in $\alpha, -\alpha$ respectively. Given $u \in U_\alpha - \{1\}$, let $v \in U_{-\alpha}$ send $u(c')$ to c, and let $v' \in U_{-\alpha}$ send c to $u^{-1}(c')$. Then vuv' switches c and c', and fixes the wall $\partial\alpha$, so by (6.3) it interchanges α with $-\alpha$. □

Definition of U_w. Let $c \in \Sigma$ be some fixed chamber, and identify W with the automorphism group of Σ. Given $w \in W$ take some reduced expression

$$w = r_{i_1} \ldots r_{i_\ell}, \text{ and set } w_o = 1, \quad w_t = r_{i_1} \ldots r_{i_t}.$$

If $\beta_j \in \Phi$ denotes the unique root of Σ containing $w_{j-1}(c)$ but not $w_j(c)$, then by (2.7) the β_j are precisely the roots containing c but not $w(c)$. We set

$$U_w = U_{\beta_1} \ldots U_{\beta_\ell}.$$

Recall that, by (2.11), any reduced expression for w can be transformed to any other by a sequence of elementary homotopies, replacing $r_i r_j \ldots (m_{ij}$ times) by $r_j r_i \ldots (m_{ij}$ times). In the sequence $\beta_1, \ldots, \beta_\ell$ this replaces a subsequence $\gamma_1, \ldots, \gamma_m$ in an $\{i, j\}$-residue by $\gamma_m, \ldots, \gamma_1$ (where $m = m_{ij}$), and so U_w will be well-defined if $U_{\gamma_1} \ldots U_{\gamma_m} = U_{\gamma_m} \ldots U_{\gamma_1}$, and this follows from the commutator relation (M2), as in (6.12)(ii). Moreover U_w is a group; this can be seen by applying (M2) again and using the fact that any pair of the roots $\beta_1, \ldots, \beta_\ell$ is prenilpotent, and for $s < t$, $[\beta_s, \beta_t]$ is a subset of $\beta_s, \ldots, \beta_t$ (Exercise 15). Furthermore the factorization of $u \in U_w = U_{\beta_1} \ldots U_{\beta_\ell}$ as $u_1 \ldots u_\ell$, where $u_t \in U_{\beta_t}$, is unique: indeed if $u = u_1' \ldots u_\ell'$ with $u_t' \in U_{\beta_t}$, then $u_2 \ldots u_\ell = u_1^{-1} u_1' \ldots u_\ell'$ fixes the chamber of $-\beta_1$ adjacent to c; this implies $u_1^{-1} u_1' = 1$, and an obvious induction does the rest.

(6.15) THEOREM. *If Δ is a Moufang building, then U_w acts simple-transitively on the set of chambers d such that $\delta(c, d) = w$. In particular if (B, N) is a Tits system on Δ, with B stabilizing c and N stabilizing Σ, then every such chamber can be written uniquely as a coset uwB where $u \in U_w$.*

PROOF: Let $w = w's$ be reduced (i.e., $\ell(w') < \ell(w)$) and let β be the unique root of Σ containing $d' = w'(c)$ but not $d = w(c)$. If x is any chamber with $\delta(c, x) = w$, let x' be the unique chamber adjacent to x with $\delta(c, x') = w'$. By induction on $\ell(w)$ there is a unique element $u \in U_{w'}$ sending d' to x'. Moreover there is a unique element $v \in U_\beta$ sending d to $u^{-1}(x)$. Clearly $uv \in U_w$ sends $d = w(c)$ to x, so U_w is transitive on $\{x \mid \delta(c, x) = w\}$, and simple-transitivity follows from the uniqueness of u and v. □

Given a Moufang building with a system of root groups $(U_\alpha)_{\alpha \in \Phi}$, let G be the group generated by the U_α. Then take N to be the subgroup generated by the $m(u)$ for $u \in U_\alpha$, as α ranges over Φ, and let H denote the subgroup of N fixing all chambers of Σ. Given some chamber $c \in \Sigma$, let Φ^+ denote the set of roots of Σ containing c, called the *positive roots*, and define

$$B = \langle H, U_\alpha \mid \alpha \in \Phi^+ \rangle.$$

(6.16) PROPOSITION. *With the notation above (B, N) is a Tits system for G, and $B \cap N = H$.*

PROOF: First notice that $G = \langle B, N \rangle$. Indeed if α is any root then either $\alpha \in \Phi^+$ in which case $U_\alpha \subset B$, or $-\alpha \in \Phi^+$ in which case for $u \in U_\alpha - \{1\}$, $U_\alpha = m(u)U_{-\alpha}m(u)^{-1} \subset \langle B, N \rangle$. Now using (5.2) it suffices to check strong transitivity. This follows from the fact that $U_w \subset B$ is transitive on chambers at distance w from c, and that by (M3) N is transitive on the chambers of Σ. □

We now set

$$U = \langle U_\alpha \mid \alpha \in \Phi^+ \rangle.$$

By (M4) H normalizes each U_α, so $B = UH$ where U is normal in B. In the spherical case $U = U_w$ where w is the longest word of W (Exercise 16), so by (6.15) U acts simple-transitively on the chambers opposite c.

(6.17) THEOREM. *For any Moufang building of spherical type, let B be a group of automorphisms containing U and fixing c, and let H be the subgroup of B fixing Σ pointwise. Then B is the semi-direct product $U \rtimes H$.*

PROOF: Since α determines U_α uniquely, H normalizes U_α for all $\alpha \in \Phi$, so U is normal in B. Moreover if c' is the chamber of Σ opposite c, then for any $g \in B$ there is a unique $u \in U$ such that $g(c') = u(c')$. Since $u^{-1}g$ fixes c and c' it fixes Σ; thus $u^{-1}g \in H$, and the uniqueness of u implies $B = U \rtimes H$. □

Example. $GL_n(k)$. In Chapter 5 we saw that the stabilizer B of a chamber is the group of upper triangular matrices. In this case H is the group of diagonal matrices, and U is the subgroup of B consisting of unipotent matrices (eigenvalues all equal to 1)

$$U = \begin{bmatrix} 1 & & * \\ & \ddots & \\ 0 & & 1 \end{bmatrix}.$$

Notice that the group generated by the U_α (for $\alpha \in \Phi$) is $SL_n(k)$, not $GL_n(k)$.

Notation. Recall from Chapter 5 that P_J is the parabolic subgroup which is the stabilizer of the face of c of type J, call it σ_J. We shall write $\Sigma_J = \Sigma \cap St(\sigma_J)$ (an apartment of $St(\sigma_J)$), and $\Phi_J = \{\alpha \in \Phi \mid \sigma_J \in \partial\alpha\}$

- see Figure 6.8, where the shaded area denotes Σ_J.

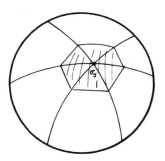

Figure 6.8

We now set

$$U_J = \langle U_\alpha \mid \Sigma_J \subset \alpha \in \Phi \rangle$$
$$L_J = \langle H, U_\alpha \mid \alpha \in \Phi_J \rangle.$$

Notice that the roots containing Σ_J are all positive, so U_J is a subgroup of U. Moreover the sets of roots for U_J and L_J are disjoint.

(6.18) THEOREM. *For a Moufang building of spherical type*

$$P_J = U_J \rtimes L_J.$$

Moreover if σ' is the simplex of Σ opposite σ_J, then L_J is the subgroup fixing σ_J and σ'.

PROOF: Every positive root either contains Σ_J or lies in Φ_J, so $B = UH \leq \langle U_J, L_J \rangle$. Therefore $\langle U_J, L_J \rangle$ is a parabolic subgroup P_K. Moreover U_J and L_J stabilize σ_J, so $K \subset J$; but L_J does not stabilize σ_K for $K \subsetneq J$, hence $K = J$, and $\langle U_J, L_J \rangle = P_J$.

We now show that L_J normalizes U_J, so let α be a root containing Σ_J. If $\beta \in \Phi_J$ then $\alpha \neq \pm\beta$, and all roots of $[\alpha, \beta]$, except β itself, contain Σ_J. By (6.12) $[U_\alpha, U_\beta] \leq U_{(\alpha,\beta)}$, and hence for $g \in U_\beta$ one has $gU_\alpha g^{-1} \subset U_\alpha U_{(\alpha,\beta)} \subset U_J$. Moreover H normalizes each U_α, and therefore L_J normalizes U_J, and $P_J = U_J L_J$.

Finally let σ' be the simplex of Σ opposite σ_J. Since $\sigma' \in \partial\beta$ for each $\beta \in \Phi_J$ (by Exercise 4), we see that L_J fixes σ'. Moreover by Exercise

17, U_J acts simple-transitively on the set of simplexes opposite σ_J, and therefore $U_J \cap L_J = 1$, and $P_J = U_J \rtimes L_J$. Moreover if $g \in P_J$, then $g = uh$ where $u \in U_J$, $h \in L_J$, and so if g fixes σ' then $u = 1$ and $g \in L_J$.

□

Example $GL_n(k)$. In Chapter 5 we gave an example of a parabolic subgroup of $GL_n(k)$ which had a $GL_m(k)$ block on the diagonal.

$$P_J = \begin{bmatrix} * & & & & & * \\ & \ddots & & & & \\ & & \boxed{GL_m(k)} & & & \\ & & & & \ddots & \\ 0 & & & & & * \end{bmatrix}$$

In this case

$$U_J = \begin{bmatrix} 1 & & & & & * \\ & \ddots & & & & \\ & & \boxed{I_m} & & & \\ & & & & \ddots & \\ 0 & & & & & 1 \end{bmatrix}$$

$$L_J = \begin{bmatrix} * & & & & & 0 \\ & \ddots & & & & \\ & & \boxed{GL_m(k)} & & & \\ & & & & \ddots & \\ 0 & & & & & * \end{bmatrix}$$

Here $L_J \cong GL_m(k) \times k^\times \times \ldots \times k^\times$, where there are $(n - m)$ copies of $k^\times$.

Notes. Much of this chapter can be found in Tits [1974]: everything in section 1 is in his Chapters 2 and 3, and the important theorem (6.6) is proved in Chapter 4; the definition of root groups and the Moufang condition for spherical buildings appears in the Addenda on pages 274-276, where the non-existence of (thick) H_3 buildings (6.10) is stated. The proof

of that result appeared later in Tits [1977]. The concept of root groups and
a Moufang condition in the general case is very recent and appears in work
of Tits [1987] on Kac-Moody groups (for an introduction to the theory of
these groups, and further references, see Tits [1985]).

Exercises to Chapter 6

1. Show that the opposition involution induces a reversal of the diagram
   in the cases of A_n and E_6. [HINT: For A_n you may use the fact that
   $W \cong S_{n+1}$ has a trivial centre; for E_6 you may use the well-known
   fact that there are exactly 27 vertices corresponding to each of the two
   end nodes].

2. Show that op_W induces a non-trivial diagram symmetry for D_n if and
   only if n is odd. [HINT: Show that the vertices of W can be regarded
   as all n-tuples whose entries are $+, -$ or 0, except those with a single
   zero: op_W switches $+$ and $-$].

3. Show that if σ and σ' are opposite simplexes, then $St(\sigma)$ and $St(\sigma')$
   are isomorphic as simplicial complexes (though as chamber systems
   they may be defined over different subsets J and J' of I).

4. Let σ and σ' be opposite simplexes of an apartment Σ. If σ lies in a
   wall M of Σ, show that σ' does too, and if σ, σ' both lie in a root γ,
   then $\sigma, \sigma' \in \partial\gamma$.

5. Given chambers x and y, in a spherical building, such that $d(x, y) =
   d - 1$ ($d = \text{diam}(W)$), show that the convex hull of x and y is a root.
   [HINT: In an apartment containing x and y, use (2.7) and (2.15) to
   count the roots containing both, then apply (2.8) and (3.8)].

6. Let Σ be an apartment of a spherical building, and let σ, σ' be opposite
   simplexes of Σ. Show that Σ is the only apartment containing $\Sigma \cap St(\sigma)$
   and σ'.

7. For any root α, let $d \in \alpha$ be a chamber having an i-panel in $\partial\alpha$, and let
   $c \in \alpha$ be j-adjacent to d, where i and j are connected in the diagram.
   Show that c has no panel in $\partial\alpha$.

8. If each connected component of a spherical diagram has rank ≥ 3,
   show that any root α contains a chamber c none of whose faces of
   codimension 1 or 2 lie in $\partial\alpha$. [HINT: Reduce to considering only
   A_3, C_3 and H_3 because in a higher rank case $\partial\alpha$ must contain a face
   of one of these types].

9. Given α and c as in Exercise 8, show that an automorphism fixing α and $E_2(c)$ must fix every chamber having a panel in $\alpha - \partial\alpha$. [HINT: Reduce to the rank 3 case].

10. Recall (from Exercise 14 of Chapter 3) that a generalized 3-gon Δ is a projective plane. A plane is called *Moufang* if for each flag (p, L) the group $U_{(p,L)}$ stabilizing each line on p, and each point on L, is transitive on the points $\neq p$ of a line $M \neq L$ on p (or equivalently the lines $\neq L$ on a point $q \neq p$ of L). Show that the Moufang condition for Δ means the same whether we treat Δ as a plane or a generalized 3-gon.

11. In Example 1 show that:

    (i) U_α is isomorphic to the additive group of k.

    (ii) U_α and $U_{-\alpha}$ generate $SL_2(k)$.

    (iii) The chambers of α are (after possibly interchanging α and $-\alpha$) the maximal flags

    $$\langle e_{\sigma(1)} \rangle \subset \langle e_{\sigma(1)}, e_{\sigma(2)} \rangle \subset \ldots \subset \langle e_{\sigma(1)}, \ldots, e_{\sigma(n)} \rangle$$

    where $\sigma(i) < \sigma(j)$.

    (iv) U_α is indeed the root group for the root α.

12. In Example 1 let c be a chamber of α having no panel in $\partial\alpha$, so U_α is the group fixing α and all chambers of $St(\pi)$ for each of the n panels π of c (see (6.5)). Show that the group fixing α and all chambers of $St(\pi)$, for only m of the panels π of c, is isomorphic to $U_\alpha \times k^\times \times \ldots \times k^\times$ ($n - m$ copies of $k^\times$).

13. Consider a spherical Coxeter complex with a connected diagram. Show that for the cases A_n, D_n, E_6, E_7, E_8, H_3, H_4, and $I_2(m)$ with m odd (see Appendix 5) W is transitive on the set of roots, and in the other cases there are two orbits. Conclude that in a Moufang building of spherical type there is only one conjugacy class of root groups in the single bond case, and two classes otherwise.

14. Let α and β be roots in an arbitrary Coxeter complex. Show that:

    (i) $\alpha \not\subset \pm\beta$ and $\beta \not\subset \pm\alpha \Leftrightarrow \partial\alpha \cap \partial\beta$ has codimension 2.

    (ii) In the spherical case $\alpha \not\subset \pm\beta \Leftrightarrow \alpha \neq \pm\beta$.

    (iii) $\{\alpha, \beta\}$ is prenilpotent $\Leftrightarrow \alpha \subset \beta$, or $\beta \subset \alpha$, or $\partial\alpha \cap \partial\beta$ has codimension 2.

15. If $\{\alpha, \beta\}$ is a prenilpotent pair of roots in a Coxeter complex, show that $[\alpha, \beta]$ is finite. Moreover let $(c_o, c_1, \ldots, c_\ell)$ be a minimal gallery, and let β_t denote the unique root containing c_t but not c_{t-1}. Then for $1 \le i \le j \le \ell$, $\{\beta_i, \beta_j\}$ is prenilpotent, and $[\beta_i, \beta_j] \subset \{\beta_t \mid i \le t \le j\}$.

16. Show that B is the normalizer of U, and in the spherical case show that $U = U_w$ where w is the longest word of W.

17. With the notation of this chapter, prove that U_J acts simple-transitively on the simplexes opposite σ_J (compare (5.4)(iv)). If τ is any such simplex what is the stabilizer of σ_J and τ? What is the normalizer of L_J in P_J?

18. Let Δ be the barycentric subdivision of the $Sp_4(k)$ quadrangle of Exercise 19 in Chapter 3. It is not a thick building, and has parameters $(1, t)$ where $t = \text{card } k$. If c and b are opposite chambers show that the subgroup fixing $E_1(c)$ and b is not the identity (cf. (6.4)).

Chapter 7
A CONSTRUCTION OF BUILDINGS

In this chapter we construct buildings which conform to a blueprint; this is the case for all Moufang buildings.

1. Blueprints.

In this section we shall introduce blueprints, and construct buildings which conform to a blueprint. We use I and M as before.

A *parameter system* will mean a collection of disjoint *parameter sets* (S_i) $i \in I$, each having a distinguished element $\infty_i \in S_i$. We shall write $S_i' = S_i - \{\infty_i\}$.

A *labelling* of a building Δ over I, *based at* $c \in \Delta$, assigns to each i-residue R a bijection

$$\phi_R \ : \ S_i \to R$$

such that $\phi_R(\infty_i) = \text{proj}_R c$. For $x \in R$, $\phi_R^{-1}(x)$ is called its *i-label*.

Example. Let S be a generalized m-gon over $\{i, j\}$, with a labelling based at $s \in S$ using the parameter system (S_i, S_j). Given any chamber $x \in S$ at distance d from s one has a gallery $(s = x_o, x_1, \ldots, x_d = x)$. If $d < m$ this is unique, and if $d = m$ there are two such galleries, of types $p(i, j)$ and $p(j, i)$. Now let u_t be the label attached to x_t in the rank 1 residue containing x_{t-1} and x_t. The gallery thus determines the sequence $(u_1, \ldots, u_d)$ where the u_t lie alternately in S_i' and S_j', and any such sequence obviously determines a unique gallery, and hence a unique chamber at the end of this gallery. If $d = m$ exactly two sequences determine the same chamber; we call these sequences *equivalent*. These equivalences, one for each chamber opposite s, give complete data for reconstructing S. $\qquad \square$

A *blueprint* is a parameter system (S_i) together with, for each distinct $i, j \in I$, a generalized m_{ij}-gon S_{ij} having a labelling by (S_i, S_j) based at some chamber $\infty_{ij} \in S_{ij}$. (In particular, in the rank 2 case a blueprint is simply a labelling of a rank 2 building).

A building Δ of type M will be said to *conform* to a blueprint if it admits a labelling by the (S_i) such that for every $\{i, j\}$-residue R there is an isomorphism $\phi_R : S_{ij} \to R$ with the property that x and $\phi_R(x)$ have the same i and j-labels for each $x \in S_{ij}$.

Now let Δ be a building having a labelling based at c, using the parameter system $(S_i)_{i \in I}$. For any chamber $x \in \Delta$ take a gallery $\gamma = (c = c_o, c_1, \ldots, c_\ell = x)$ of reduced type $f = i_1 \ldots i_\ell$ from c to x (where $r_f = \delta(c, x)$ of course). Then γ determines a sequence $(u_1, \ldots, u_\ell)$, where $u_t \in S'_{i_t}$ is the i_t-label of c_t. Conversely such a sequence determines a gallery γ starting at c, and hence a chamber x at the end of this gallery.

Now suppose Δ conforms to a blueprint $(S_i, S_{ij})_{i,j \in I}$. If f' is elementary homotopic to f, then we have a gallery $\gamma' = \gamma_1 \omega' \gamma_2$ of type f' from c to x, where $\gamma = \gamma_1 \omega \gamma_2$ and ω, ω' are galleries in an $\{i, j\}$-residue, corresponding to sequences which are equivalent in S_{ij}. Thus using the blueprint, and concatenating elementary homotopies, we can transform one sequence $(u_1, \ldots, u_\ell)$ to another. The chambers of Δ could then be defined as equivalence classes of such sequences, but there is a problem. Transforming $(u_1, \ldots, u_\ell)$ to another sequence of the same type f should give the same sequence; this means we must consider what happens to $(u_1, \ldots, u_\ell)$ when we apply a self-homotopy of f. By (2.17) self-homotopies are generated in rank 3 spherical residues, and so this leads to the following theorem, in which we call a blueprint *realisable* if there is a building which conforms to it.

(7.1) THEOREM. *A blueprint is realisable if its restriction to each spherical rank 3 subdiagram is realisable. In this case there is a unique building which conforms to it.*

The Construction. Given a blueprint we first construct a chamber system S as follows. The chambers of S are sequences

$$\overline{u} = (u_1, \ldots, u_\ell)$$

where $u_t \in S'_{i_t}$ and $f = i_1 \ldots i_\ell$ is reduced. We call f the *type* of $\overline{u}$. We define i-adjacency via

$$(u_1, \ldots, u_\ell) \underset{i}{\sim} (u_1, \ldots, u_\ell, u_{\ell+1}) \underset{i}{\sim} (u_1, \ldots, u_\ell, u'_{\ell+1}),$$

if $u_{\ell+1}, u'_{\ell+1} \in S_i$; this is evidently an equivalence relation, so S is a chamber system.

We define an *elementary equivalence* to be an alteration from a sequence $\bar{u}_1 \bar{u} \bar{u}_2$ of type $f_1 p(i,j) f_2$ to $\bar{u}_1 \bar{u}' \bar{u}_2$ of type $f_1 p(j,i) f_2$ where $\bar{u}$ and $\bar{u}'$ are equivalent in S_{ij}. Two sequences $\bar{u}$ and $\bar{v}$ are called *equivalent*, written $\bar{u} \simeq \bar{v}$, if one can be transformed to the other by a sequence of elementary equivalences.

The chamber system we want is S/equivalence; we call it C. Its chambers are equivalence classes of sequences $\bar{u} = (u_1, \ldots, u_\ell)$, denoted $[\bar{u}]$ or $[u_1, \ldots, u_\ell]$. Notice that $[\bar{u}]$ determines a unique element $r_f \in W$ where f is the type of $\bar{u}$; we call this $\rho[\bar{u}]$. We now define i-adjacency in C by $x \underset{i}{\sim} y$ if $x = [\bar{u}]$, $y = [\bar{v}]$ with $\bar{u} \underset{i}{\sim} \bar{v}$; in Step 2 of the proof below we shall see that this is in fact an equivalence relation.

PROOF OF THEOREM (7.1): We show that C is a building conforming to the given blueprint.

Step 1. If $\bar{u} \simeq \bar{v}$, and $\bar{u}, \bar{v}$ both have type f, then $\bar{u} = \bar{v}$.

Certainly the equivalence $\bar{u} \simeq \bar{v}$ induces a self-homotopy of f, and by (2.17) we need only consider equivalences $\bar{w} \simeq \bar{w}'$ giving self-homotopies which are either inessential or lie in rank 3 spherical residues. The former case is easily seen to imply $\bar{w} = \bar{w}'$, and the latter case does too because of our rank 3 hypothesis.

Step 2. i-adjacency is an equivalence relation.

Let $x \underset{i}{\sim} y \underset{i}{\sim} z$. If $\rho(x)$, $\rho(y)$ and $\rho(z)$ are not i-reduced on the right, then we have $x = [\bar{u}, u']$, $y = [\bar{u}, u] = [\bar{v}, v]$, $z = [\bar{v}, v']$, where $\bar{u}, \bar{v}$ have types f, g respectively, and $u', u, v, v' \in S_i'$. Evidently $fi \simeq gi$, so $f \simeq g$, and $\bar{u} \simeq \bar{v}_o$ for some $\bar{v}_o$ of type g. Therefore $(\bar{u}, u) \simeq (\bar{v}_o, u)$ of type gi, so by Step 1 $\bar{v}_o = \bar{v}$ and $u = v$. Therefore $x = [\bar{v}, u'] \underset{i}{\sim} [\bar{v}, v'] = z$ as required. A similar proof works if one of $\rho(x)$, $\rho(y)$, $\rho(z)$ is i-reduced on the right; one simply deletes u', u, v, v' as appropriate.

Step 3. C is a chamber system satisfying (P_c) where $c = [\emptyset]$ [Recall (P_c): Given two galleries starting at c and ending at the same chamber, of reduced types f and f', one has $r_f = r_{f'}$].

C is a chamber system by Step 2, and it is straightforward to see, by induction on the length of f, that a gallery $(c = c_1, c_2, \ldots, c_\ell = d)$ of reduced type f corresponds to a sequence $\bar{u}$ of type f such that $[\bar{u}] = d$. Now if $\bar{u}'$ has type f' and $\bar{u} \simeq \bar{u}'$, then $f \simeq f'$, so $r_f = r_{f'}$.

Step 4. C is a chamber system of type M.

Let R be any $\{i,j\}$-residue, and $x = [\overline{u}, \overline{v}]$ any chamber of R, where the type of $\overline{u}$ is $\{i,j\}$-reduced on the right, and the type of $\overline{v}$ involves only i and j. If $x \underset{i}{\sim} y$, then $y = [\overline{u}, \overline{v}']$ where $\overline{v}' \underset{i}{\sim} \overline{v}$, so by $\{i,j\}$-connectivity of R each of its chambers has the form $[\overline{u}, \overline{w}]$ where the type of $\overline{w}$ involves only i and j. Moreover any such chamber lies in R, so the map $[\overline{u}, \overline{w}] \to [\overline{w}]$ from R to S_{ij} is surjective, and by Step 1 it is an isomorphism. Thus $R \cong S_{ij}$, as required.

Finally from Steps 3 and 4 it follows, by (4.2), that C is a building. Moreover C obviously acquires a labelling conforming to the blueprint, and its uniqueness is an immediate consequence of this. $\square$

Remark. We could have defined i-adjacency as the equivalence relation generated by the i-adjacency we in fact defined. In this case Step 2 would have to be rephrased by saying that if $x \underset{i}{\sim} y$ then we can write $x = [\overline{u}, u]$, $y = [\overline{u}, v]$ where u, v are either non-existent or contained in S_i' .

Before leaving this section we state a theorem which follows immediately from the proof of (7.1).

(7.2) THEOREM. *A blueprint is realisable if and only if for any two sequences $\overline{u}, \overline{v}$ of the same reduced type, $\overline{u} \simeq \overline{v}$ implies $\overline{u} = \overline{v}$.*

PROOF: The hypothesis is Step 1 of the proof of (7.1), so the remainder of the proof of (7.1) goes through unchanged. The "only if" part follows from the fact that a sequence $\overline{u}$ of type f corresponds to a gallery of type f from $c = [\emptyset]$ to $[\overline{u}]$, and such galleries are unique (3.1)(v). $\square$

2. Natural Labellings of Moufang Buildings.

In Chapter 6 section 4 we defined Moufang buildings. If Δ is Moufang, and Φ is the set of roots in an apartment Σ, then there is a set of root groups U_α, one for each $\alpha \in \Phi$, having certain properties (M1) - (M4). All spherical buildings are Moufang if the diagram has no connected component of rank ≤ 2. In this section we show that a Moufang building conforms to a blueprint, in fact a "natural" blueprint defined using root group elements.

Choose a chamber $c \in \Sigma$, let π_i be its panel of type i, and define $\alpha_i \in \Phi$ to be the root containing c and with $\pi_i \in \partial \alpha_i$. We let r_i denote the reflection interchanging α_i and $-\alpha_i$ and write $U_i = U_{\alpha_i}$. Now for each $i \in I$ select some element $e_i \in U_i - \{1\}$. Recall from Chapter 6 that for each $u \in U_\alpha - \{1\}$ there is a unique element $m(u) \in U_{-\alpha} u U_{-\alpha} \cap N$ interchanging α and $-\alpha$.

(7.3) LEMMA. *Setting $n_i = m(e_i)$ one has*

$$n_i n_j \ldots = n_j n_i \ldots (m_{ij} \text{ terms alternating } n_i \text{ and } n_j \text{ on each side}).$$

Consequently for any $w \in W$, there is a unique $n(w) \in N$, where

$$n(w) = n_{i_1} \ldots n_{i_\ell} \quad \text{for } w = r_{i_1} \ldots r_{i_\ell} \text{ (reduced)}.$$

PROOF: The first statement is proved in Appendix 1, and the second statement is an immediate consequence of the first, using (2.11).   □

In Chapter 6 section 4 we defined a group U_w for each $w \in W$. It acts simple-transitively on the set of chambers d such that $\delta(c,d) = w$, and if B denotes the stabilizer of c, then by (6.15) every such d can be represented in a unique way as a coset uwB, where $u \in U_w$. Since $wB = n(w)B$, we have a bijection between chambers of Δ and elements $un(w)$, where $u \in U_w$. The fact that we are able to omit B (which is in general a complicated group) means that the structure of the building is remarkably simple; in fact it conforms to a blueprint.

(7.4) LEMMA. *If $w = r_f$ for some reduced word $f = i_1 \ldots i_k$, then for $u \in U_w$,*

$$un(w) = u_1 n_{i_1} \ldots u_k n_{i_k}$$

where $u_t \in U_{i_t}$, and this factorisation is unique.

PROOF: Let $w' = r_g$ where $g = i_2 \ldots i_k$, so $w = r_{i_1} w'$. If $\beta_1, \ldots, \beta_k$ are the roots separating c from $w(c)$ in Σ, their order determined by the gallery of type f from 1 to w, then $u = v_1 \ldots v_k$ where $v_t \in U_{\beta_t}$ (see 6.15). Therefore

$$un(w) = v_1 \ldots v_k n(w)$$
$$= v_1 n_{i_1} n_{i_1}^{-1} v_2 \ldots v_k n_{i_1} n(w')$$
$$= v_1 n_{i_1} vn(w')$$

where $v \in U_{w'}$, because n_{i_1} switches $\beta_2, \ldots, \beta_k$ with the roots separating c from $w'(c)$, their order determined by the gallery of type g from 1 to w'. The factorisation now follows by induction on the length of w, and its uniqueness follows from the uniqueness of the decomposition $u = v_1 \ldots v_k$ which is a consequence of (6.15).   □

The Natural Labelling given by the $(e_i)_{i \in I}$. The lemma above implies that each chamber of Δ can be written as an equivalence class $un(w)$ of elements of the form $u_i n_{i_1} \ldots u_k n_{i_k}$ having type $f = i_1 \ldots i_k$ where $r_f = w$. It is this which gives what we call a *natural labelling* of the building Δ. More precisely let R be any i-residue of Δ, and let $\text{proj}_R c = d$ and $w = \delta(c, d)$. As cosets of B the chambers of R may be written $un(w)B$ (this is d), and $un(w)vn_iB$ where $u \in U_w$ and $v \in U_i$. We assign them the i-labels ∞_i and v, using $S_i = U_i \cup \{\infty_i\}$. If we let S_{ij} be the $\{i, j\}$-residue containing c, then S_{ij} acquires a labelling and we have a blueprint given by the $(e_i)_{i \in I}$.

(7.5) PROPOSITION. *The natural labelling of Δ above conforms to the blueprint given by its restriction to $E_2(c)$. In particular a Moufang building conforms to a blueprint.*

PROOF: If A is any $\{i, j\}$-residue, let $w = \delta(c, \text{proj}_A c)$. As a coset of B, $\text{proj}_A c$ is $un(w)B$ for some $u \in U_w$, and left multiplication by $un(w)$ gives an isomorphism from the $\{i, j\}$-residue containing c to A, preserving i and j-labels.   □

(7.6) LEMMA. *A Moufang plane has a unique natural labelling in the sense that any natural labelling can be transformed to any other by an automorphism of the plane fixing the base chamber.*

PROOF: We shall not prove this here: a proof is given in Ronan-Tits [1987] Lemma 2.   □

We now extend the concept of a natural labelling to generalized 2-gons, by defining a labelling using (S_1, S_2) to be *natural* if (u_1, u_2) is equivalent to (u_2, u_1) for any $u_1 \in S_1'$, $u_2 \in S_2'$. If Δ is a Moufang building with a natural labelling given by $e_i \in U_i - \{1\}$ then any $A_1 \times A_1$ residue acquires a natural labelling in this sense (because the appropriate root groups commute - see Exercise 1).

Finally we remark that if Δ is a direct product $\Delta_1 \times \ldots \times \Delta_r$, then labellings of the Δ_j generate a labelling of Δ in an obvious way: if Δ_j is over I_j (so Δ is over $\cup I_j$), and if $i \in I_j$ then the chamber $(c_1, \ldots, c_r)$ of $\Delta_1 \times \ldots \times \Delta_r$ has the same i-label as c_j in Δ_j. If $\Delta = \Delta_1 \times \Delta_2$ is an $A_1 \times A_1$ building, this gives what we have called a natural labelling.

3. Foundations.

Take a parameter system $(S_i)_{i \in I}$, and for each $i, j \in I$ a generalized m_{ij}-gon S_{ij} (not labelled) with a base chamber ∞_{ij}. Let

$$\phi_{ij} : S_i \to S_{ij}$$

be a bijection onto the i-residue of S_{ij} containing ∞_{ij}, and sending ∞_i to ∞_{ij}. A *foundation* of type M is the amalgamated sum of the S_{ij} with respect to the ϕ_{ij}; in other words the union of the S_{ij} with the identifications $\phi_{ij}(s_i) = \phi_{ik}(s_i)$ for all $s_i \in S_i$, and for all $i, j, k \in I$. It is a chamber system E over I having a base chamber c identified with all ∞_i and ∞_{ij}, and is the union of the rank 2 residues containing c. We say E *supports* a building Δ if it is isomorphic to the union of the rank 2 residues of Δ containing some given chamber c of Δ (i.e. $E_2(c)$).

A *labelling* of E is defined in the obvious way: if π is the i-panel of c, then $St(\pi) = S_i$, and if π is any other i-panel one takes a bijection $S_i \leftrightarrow St(\pi)$ such that ∞_i corresponds to the chamber of $St(\pi)$ nearest the base chamber c. Notice that a labelling of a foundation is nothing other than a blueprint.

(7.7) LEMMA. *Let E be a rank 3 foundation of reducible type (i.e. disconnected diagram). Then E supports a building Δ which is uniquely determined up to isomorphism, and Δ conforms to any labelling of E whose restriction to $A_1 \times A_1$ residues is natural.*

PROOF: Let $I = \{1, 2, 3\}$ with $m_{12} = m_{13} = 2$. By (3.10) any such building is a direct product $\Delta_1 \times \Delta_{23}$, so Δ must be $S_1 \times S_{23}$ with the labelling generated as above. □

(7.8) PROPOSITION. *Let E be an A_3 or C_3 foundation which supports a building Δ. Then Δ conforms to any labelling of E whose restriction to the rank 2 residues is natural, and Δ is uniquely determined up to isomorphism by E.* □

PROOF: Let $I = \{1, 2, 3\}$ with $m_{12} = 3$ and $m_{13} = 2$, and let $\mathcal{L}$ be a labelling of E whose restrictions $\mathcal{L}_{ij}$ to S_{ij} are natural. By (6.7) Δ is Moufang, and hence conforms to a natural labelling $\mathcal{L}'$ of E extending $\mathcal{L}_{23}$ (given e_2, e_3 choose any e_1). By (7.6) there is an automorphism θ of S_{12} fixing S_2 and carrying $\mathcal{L}'_{12}$ to $\mathcal{L}_{12}$, and we extend θ to an automorphism of E which is the identity on S_{23} (and hence fixes $\mathcal{L}_{23}$). Since $\mathcal{L}_{13}$ is uniquely determined by its restrictions to S_1 and S_3, we see that θ sends $\mathcal{L}'_{13}$ to $\mathcal{L}_{13}$, and hence sends $\mathcal{L}'$ to $\mathcal{L}$. Thus Δ conforms to $\mathcal{L}$, as required. Moreover, if Δ is any other building supported by E, it too conforms to the given labelling of E, and is therefore isomorphic to Δ. □

(7.9) THEOREM. *Let E be a foundation with no residue of type H_3. If each residue of type A_3 or C_3 supports a building, then E supports a building. If E is of spherical type this building is uniquely determined up to isomorphism.*

PROOF: Choose a natural labelling for each S_{ij} whenever $\{i,j\}$ is in an A_3 or C_3 residue, or is of type $A_1 \times A_1$ in a spherical triple. Choose other labellings arbitrarily. This gives a blueprint which by (7.7) and (7.8) is realisable for rank 3 spherical residues, and hence by (7.1) there is a building Δ which conforms to it.

To prove uniqueness it suffices to consider the case of a connected diagram, since Δ is a direct product of buildings for the connected components of the diagram. In this case we may assume a diagram of rank ≥ 3, and since E is of spherical type, Δ is Moufang and the diagram has at most one double bond. We apply the technique in the proof of (7.8): let $\mathcal{L}$ be a labelling of E whose restrictions $\mathcal{L}_{ij}$ to S_{ij} are natural. If $x, y \in I$ are the nodes of the double bond (or any two nodes if no double bond exists), then Δ conforms to a natural labelling $\mathcal{L}'$ extending $\mathcal{L}_{xy}$ (given e_x, e_y choose the other e_i arbitrarily). As in the proof of (7.8), (7.6) allows us to define an isomorphism of E sending $\mathcal{L}'$ to $\mathcal{L}$. Thus Δ conforms to $\mathcal{L}$, and is therefore unique up to isomorphism. □

Remark. In the next chapter we shall deal with the case of A_3 and C_3 blueprints (A_3 in detail, but C_3 only by using Tits' classification [1974]). When we have done so it will be quite clear that buildings exist for all possible diagrams which have no H_3 subdiagram. However, the reader certainly has enough information at the moment to deal with many cases (see Exercises 3 and 4).

Notes. Everything in this chapter appears in Ronan-Tits [1987], except that "BN-Pairs with a splitting" appear there in place of Moufang buildings. These BN-Pairs have root groups U_α for $\alpha \in \Phi$, which satisfy conditions similar to (M1) - (M4) of Chapter 6, although (M2) is weakened.

Exercises to Chapter 7

1. Let U_1 and U_2 be the fundamental root groups in an $A_1 \times A_1$ residue of a Moufang building. Given $u_1, v_1 \in U_1$ and $u_2, v_2 \in U_2$ with $u_1 n_1 u_2 n_2 = v_2 n_2 v_1 n_1$, show that $u_1 = v_1$ and $u_2 = v_2$. [HINT: $[U_1, U_2] = 1$ by (M2), and for $x \in U_1$ (or U_2), $m(x)$ commutes with U_2 (or U_1); use (7.3)].

2. If $\Delta = \Delta_1 \times \ldots \times \Delta_r$ and each Δ_j has a labelling conforming to a blueprint, show that the labelling generated on Δ conforms to a blueprint.

3. Using the existence of buildings of type $A_n(\circ\!\!-\!\!-\!\circ\!\!-\!\!\!-\ldots\!\!-\!\!-\circ)$ and of generalized m-gons for all m (see Exercise 17 of Chapter 3), prove the existence of buildings of type $\circ\!\!-\!\!\!-\ldots\!\!-\!\!-\circ\!\!-\!\!\overset{m}{-}\!\!-\circ$ for any $m > 5$.

4. Try Exercise 3 for some other diagrams, and find some diagram for which existence of a suitable foundation cannot be inferred using the results of this chapter.

Chapter 8
THE CLASSIFICATION OF SPHERICAL BUILDINGS

This chapter deals with the classification and existence of buildings of spherical type for which each connected component of the diagram has rank at least three. According to Theorem (7.9) of the preceding chapter the buildings of spherical type M are uniquely determined by foundations of type M (this is also a consequence of Theorem (6.6) in Chapter 6), and such foundations support buildings when their A_3 and C_3 residues do. Therefore the first thing we shall do here is to examine A_3 foundations.

1. A_3 Blueprints and Foundations.

Since an A_3 building is Moufang, we know by (7.7) that it conforms to a blueprint whose rank 2 restrictions are natural. We therefore need to know what the natural labelling of a Moufang plane looks like. The details are given in Appendix 1, and the main points are as follows.

The three positive root groups U_1, U_{12} and U_2 (in a natural order) are abelian (by (6.13)) and may be identified with an abelian group A written additively. Moreover, a natural labelling is determined by non-identity elements $e_1 \in U_1$ and $e_2 \in U_2$, and the identification can be done in such a way that e_1, e_2, and $e_{12} = [e_1, e_2]$ are identified with a common element $e \in A$. Using subscripts to denote membership of U_1, U_{12} or U_2 one has a multiplicative structure on A defined via $(xy)_{12} = [x_1, y_2]$ (again see Appendix 1 for details). With this addition and multiplication A becomes an alternative division algebra in which e plays the role of multiplicative identity. We mention in passing that such an algebra is either a field (not necessarily commutative), in which case the plane is Desarguesian; or, by the Bruck-Kleinfeld theorem [1951], it is a Cayley-Dickson algebra, 8-dimensional over its centre, in which case the plane is sometimes called a *Cayley plane*.

For the purposes of this section, we need the fact that the natural labelling is given by setting the following two sequences equivalent:

sequence	type
$x\ \ y\ \ z$	1 2 1
$z\ \ y'\ \ x$	2 1 2

where $y + y' = -(xz)_1$.

The subscript is needed because if we interchange the roles of 1 and 2, then we obtain the opposite algebra structure (see Appendix 1 section 2, or use the uniqueness of the natural labelling); thus

$$(xz)_1 = (zx)_2.$$

In fact we shall think of $(xz)_1$ as referring to the algebra structure induced on U_1, so in other words U_1 with its algebra structure is identified with the opposite of U_2.

We now return to the subject of A_3 blueprints. For an A_3 blueprint to be realizable it is necessary and sufficient, by (7.2), that an equivalence between two sequences of reduced type f is an equality. If the corresponding self-homotopy of f is inessential then one certainly gets equality. Moreover the only essential self-homotopy is obtained from the longest word, by working around an apartment as shown below.

		sequence						type			
a	b	c	d	e	f	1	2	3	1	2	1
a	b	d	c	e	f	1	2	1	3	2	1
d	b'	a	c	e	f	2	1	2	3	2	1
d	b'	e	c'	a	f	2	1	3	2	3	1
d	e	b'	c'	f	a	2	3	1	2	1	3
d	e	f	c''	b'	a	2	3	2	1	2	3
f	e'	d	c''	b'	a	3	2	3	1	2	3
f	e'	c''	d	b'	a	3	2	1	3	2	3
f	e'	c''	a	b''	d	3	2	1	2	3	2
f	a	c'''	e'	b''	d	3	1	2	1	3	2
a	f	c'''	b''	e'	d	1	3	2	3	1	2
a	b''	c''''	f	e'	d	1	2	3	2	1	2
a	b''	c''''	d	e''	f	1	2	3	1	2	1

This self-homotopy is an equality if and only if $b'' = b$, $c'''' = c$, and $e'' = e$. We now compute using the multiplication $(xy)_1$ or $(xy)_3$, but not $(xy)_2$ as it may (and in fact does) depend on which A_2 residue it is induced from. One has

$$b + b' = -(ad)_1, \text{ and } b' + b'' = -(da)_3,$$

so $b'' = b$ if and only if

$$(xy)_1 = (yx)_3. \tag{1}$$

Given this equality we find that $e = e''$. Similarly:

$$c + c' = -(ea)_3 = -(ae)_1,$$
$$c' + c'' = -(b'f)_1,$$
$$c'' + c''' = -(ae')_1,$$
$$c''' + c'''' = -(fb)_3 = -(bf)_1.$$

Using $e' = -(fd)_3 - e = -(df)_1 - e$, and deleting the subscript 1, one obtains:

$$c - c'''' = a(df) - (ad)f.$$

So $c = c''''$ if and only if

$$x(yz) = (xy)z. \tag{2}$$

Thus we find that our blueprint is realizable if and only if equations (1) and (2) are satisfied. Equation (2) gives the well-known result that the coordinate ring is a field, so each plane is Desarguesian. Moreover since the ring structure induced (after one has chosen a unit element) on U_2 from the $\{1,2\}$-plane is opposite that induced on U_1 (see above), and similarly with 1 replaced by 3, we see from equation (1) that the blueprint is realizable if and only if the two planes induce opposite field structures on U_2.

   We rephrase this as a theorem.

(8.1) THEOREM. *An A_3 foundation E supports a building if and only if the two planes are Desarguesian and induce opposite field structures on their common punctured rank 1 residue (i.e. with the base chamber removed).*

PROOF: If E supports a building then by (7.8) this building conforms to a labelling of E of the type investigated above. On the other hand such a blueprint is realizable by (7.2).   □

2. Diagrams with Single Bonds.

   The connected spherical diagrams with single bonds are A_n, D_n, E_6, E_7, E_8.

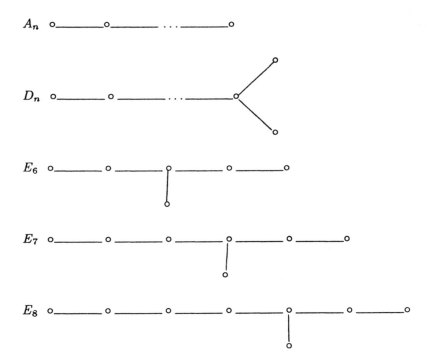

Using the results above, we have the following classification.

Type A_n. By the A_3 result the fundamental root groups $U_1, \ldots, U_n$ acquire field structures, and for $i = 2, \ldots, n-1$ the structure induced on U_i by the $\{i-1, i\}$-residue is opposite that induced by the $\{i, i+1\}$-residue. For each field k (not necessarily commutative) there is a unique foundation (up to isomorphism) and therefore by (7.9) a unique A_n building, and vice versa. This $A_n(k)$ building is the flag complex of projective space, exhibited in Example 4 of Chapter 1.

Type D_4.

Here the field structures induced on U_0 by the three types of residual planes are mutually opposite. Therefore the field is commutative (an alternative proof of this fact is given by Tits [1974] (6.12), and see also Exercise 9).

Types D_n, E_6, E_7, E_8. By the A_3 result and the D_4 result above, each root group acquires the structure of a commutative field, and these are mutually isomorphic. Therefore for each commutative field k there is a unique such foundation (up to isomorphism) giving a unique building, and vice versa.

The $D_n(k)$ Building. In the D_n case the building can be obtained as follows. Take a $2n$-dimensional vector space over k with basis $x_1, \ldots, x_n$, $y_1, \ldots, y_n$. Define the quadratic form $Q(v) = \Sigma a_i b_i$ where $v = \Sigma a_i x_i + \Sigma b_i y_i$. There are totally singular subspaces S (i.e. $Q(s) = 0 \; \forall s \in S$) of dimensions 1 up to n, those of dimension $n-1$ being contained in exactly two of dimension n (Exercise 1). For this reason the totally singular subspaces do not give a thick C_n building. However, the following construction gives a thick D_n building (see Exercises 2-4).

The chambers of the building are nested sequences of totally singular subspaces of the form

$$S_1 \subset S_2 \subset \ldots \subset S_{n-2} \; \begin{matrix} \nearrow S_n \\ \searrow S_n' \end{matrix}$$

where those of dimension $n-1$ have been omitted, and $\dim(S_n \cap S_n') = n-1$. Such sequences are called *oriflammes* in [loc. cit.] (7.12); two are adjacent if they differ in at most one term. Considering the building as a simplicial complex its vertices are all totally singular subspaces of dimension $\neq n-1$; two vertices are joined by an edge if and only if, as subspaces, one contains the other, or they both have dimension n and intersect in dimension $n-1$.

If k is an algebraically closed field, the appropriate group is the orthogonal group $O_{2n}(k)$. If k is a finite field F_q this is usually written $O_{2n}^+(q)$ to distinguish it from the other orthogonal group $O_{2n}^-(q)$, also known as $^2D_n(q)$, whose building has type C_{n-1}.

3. C_3 Foundations.

Since a C_3 building is Moufang, its A_2 and C_2 residues are Moufang (see (6.8)), and, as mentioned earlier, a Moufang plane is either Desarguesian, or is a Cayley plane. We shall treat these two cases separately.

The Case of Desarguesian planes. In this case the quadrangle has to be of classical type (cf. Exercise 8). This means it arises, as in section 4, from a hermitian or a pseudo-quadratic form of Witt index 2 on a vector space V over a field K (not necessarily commutative). The vertices (points and lines) of the quadrangle are the totally isotropic (or singular) 1- and 2-spaces of V respectively. The residues for the 2-spaces will be called line-residues because their chambers correspond to the points of a projective line (1-spaces in a 2-space) over K.

In fact the quadrangle induces a field structure K on its punctured line-residues, and in the spirit of (8.1) we can now state the following consequence of the classification in [loc. cit.] Chapter 8.

(8.2) THEOREM. *A C_3 foundation whose plane is Desarguesian supports a building if and only if the plane and the quadrangle induce oppposite field structures on their common punctured rank 1 residue.*   □

Remarks.

1. It can happen that both types of residues in the quadrangle can be taken as line-residues, namely when the dual (interchanging roles of points and lines) also arises from the 1 and 2-spaces of a vector space; these cases are shown in section 5 when we deal with the Tits diagram for a simple algebraic group.

2. In one of these special cases where both residues can be taken as line residues (the D_4/A_1^2 case), one residue acquires a canonical pair of opposite quaternion structures. In all other cases the field structure is canonical (again see section 5).

The Case of Cayley planes. A non-Desarguesian, Moufang plane induces a Cayley algebra K (8 dimensional over a commutative field k) on its punctured rank 1 residues. The quadrangle then has to arise from a 12-dimensional vector space $K \oplus k^4$ with quadratic form $n_K(x_o) - x_1 x_3 + x_2 x_4$, where n_K is the norm form of the Cayley algebra. Moreover it is the point-residue (as opposed to the line-residue) which the quadrangle has in common with the plane (a diagrammatic illustration for this is given in section 5).

(8.3) THEOREM. *A C_3 foundation whose plane is a Cayley plane supports a building if and only if the quadrangle arises from a 12-space as above, and the plane and the quadrangle induce the same proportionality class of 8-dimensional anisotropic forms on their common punctured residue (one form from the Cayley algebra, the other from the quadratic form on $W^\perp/W$ where W is a totally singular 2-space of the 12-space).* □

4. C_n Buildings for $n \geq 4$.

$$\circ\!\!-\!\!\!-\!\!\!-\!\!\circ\!\!-\!\!\!-\!\!\!-\cdots\!-\!\!\!-\!\!\circ\!\!=\!\!=\!\!\circ$$

Given a C_n diagram with $n \geq 4$, as shown, there is an A_3 subdiagram, and by (8.1) this forces the planes to be Desarguesian. In fact these C_n buildings are classified by their C_3 residues, as the following theorem makes clear.

(8.4) THEOREM. *A C_3 foundation whose plane is Desarguesian and which supports a building, extends to a unique C_n foundation supporting a unique building, and for $n \geq 4$ every C_n building arises in this way.*

PROOF: This is an immediate consequence of Theorems (7.9) and (8.1).□

As is shown in [loc. cit.] Chapter 8, all such C_n buildings can be obtained using a vector space endowed with a hermitian or pseudo-quadratic form of Witt index n. Here I shall simply explain the terminology, details being available in [loc. cit.].

Let K be a field (not necessarily commutative), σ an anti-automorphism of K with $\sigma^2 = $ id., and let $\epsilon = \pm 1$. Define

$$K_{\sigma,\epsilon} = \{t - \epsilon t^\sigma \mid t \in K\}.$$

Now let V be a right vector space over K (not necessarily finite dimensional), and let $f : V \times V \to K$ satisfy:

(0) $f(xa, yb) = a^\sigma f(x,y) b$ for all $x, y \in V$ and $a, b \in K$.

(1) $f(y, x) = \epsilon f(x, y)^\sigma$.

(2) $f(x, x) = 0$ if $\sigma = $ id. and $\epsilon = -1$, in which case f is called a *symplectic* (or *alternating*) *form*.

Condition (0) means f is a *sesquilinear* ("$1\frac{1}{2}$-linear") form relative to σ, and condition (1) implies in particular that the relationship $x \perp y$ (meaning $f(x,y) = 0$) is symmetric. Such a form will generally be called *hermitian*,

or more precisely (σ, ϵ)-*hermitian*. If $\sigma = $ id. and $\epsilon = 1$ it is often called a
symmetric bilinear form.

We now define $q : V \rightarrow K/K_{\sigma,\epsilon}$ to be a *pseudo-quadratic* form associated to f if:

(3) $q(xa) = a^\sigma q(x)a$ for all $x \in V$ and $a \in K$.

(4) $q(x + y) = q(x) + q(y) + f(x, y) + K_{\sigma,\epsilon}$ for $x, y \in V$.

When $\sigma = $ id. and $\epsilon = 1$ one has $K_{\sigma,\epsilon} = 0$ and q is called a *quadratic* form.

Notice that a non-zero sesquilinear form must map *onto* K, so q determines f except when $K = K_{\sigma,\epsilon}$. In fact $K = K_{\sigma,\epsilon}$ if and only if $\sigma = $ id. and $\epsilon \neq 1$ (Exercise 5), in which case char $K \neq 2$ (because $\epsilon \neq 1$) and f is a symplectic form.

Notice also that q is uniquely determined by its associated sesquilinear form f when char $K \neq 2$, because $q(x) = \frac{1}{2}f(x, x) + K_{\sigma,\epsilon}$. More generally if there is an element λ in the centre of K such that $\lambda + \lambda^\sigma = 1$, then $q(x) = \lambda f(x, x) + K_{\sigma,\epsilon}$ (see Exercise 6); this occurs for char $K = 2$ when the restriction of σ to the centre of K is not the identity (the ^2A_n case). A necessary and sufficient condition for f to determine q is given in [loc. cit.] 8.2.4.

A subspace W of V is called *totally isotropic* for f if $f(x, y) = 0$ for all $x, y \in W$, and *totally singular* for q if $q(x) = 0$ for all $x \in W$. All maximal totally isotropic (or totally singular) subspaces have the same dimension, called the *Witt index* (see e.g. Artin [1957] 3.10). Notice that the subspace $V^\perp = \{x \in V \mid f(x, V) = 0\}$ is totally isotropic; we call f *non-degenerate* if $V^\perp = 0$. A pseudo-quadratic form q is called *non-degenerate* if $V^\perp$ (for the associated f) has no non-zero singular vectors (i.e., $V^\perp \cap q^{-1}(0) = 0$).

We say f is *trace-valued* if

$$f(x, x) = a + \epsilon a^\sigma \text{ for some } a \in K.$$

When there are totally isotropic subspaces not contained in $V^\perp$, the property of being trace-valued is equivalent to the property that the totally isotropic subspaces span V ([loc. cit.] 8.1.6). If f arises from a pseudo-quadratic form, then it must be trace-valued (see Exercise 7 for a proof). Moreover if f is trace-valued and is not a symplectic form in odd characteristic, then it must arise from a pseudo-quadratic form.

The Building. If f is non-degenerate and trace-valued, or if q is a non-degenerate pseudo-quadratic form, of Witt index n, then the totally isotropic (t.i.), or totally singular (t.s.), subspaces determine a building of type C_n. The chambers are all maximal nested sequences

$$S_1 \subset \ldots \subset S_n$$

of t.i., or t.s., subspaces, and the other simplexes are subsequences of these (see Example 5 in Chapter 1). In particular the vertices are the t.i., or t.s., subspaces themselves. This building is thick providing the form is not the one mentioned in section 2 giving a D_n building; in that special case each t.s. $(n-1)$-space lies in exactly two t.s. n-spaces.

Theorem (8.2) is a consequence of the following theorem [loc. cit.] (8.22).

(8.5) THEOREM. *Every C_3 building whose planes are Desarguesian, and every C_n building for $n \geq 4$ arises from a non-degenerate pseudo-quadratic form, or a non-degenerate hermitian form of Witt index n.*   □

We emphasize that this vector space could be infinite dimensional; indeed its dimension might not even be countable. For example let Z be any set and let X be the disjoint union of Z and $\{x_1, \ldots, x_n, y_1, \ldots, y_n\}$. We let V denote the real vector space with basis X, whose vectors are all $v = a_1 x_1 + \ldots + a_n x_n + b_1 y_1 + \ldots + b_n y_n + \sum_{i=1}^{m} c_i z_i$ where $z_i \in Z$ and $a_i, b_i, c_i \in \mathbf{R}$. We define $q(v) = a_1 b_1 + \ldots + a_n b_n + \sum_{i=1}^{m} c_i^2$. This quadratic form has Witt index n, and if Z is non-empty we obtain a thick C_n building.

5. Tits Diagrams and F_4 Buildings.

To classify F_4 buildings one needs to know which Moufang quadrangles have the property that they and their duals arise from a form (of Witt index 2) on some vector space. It is then a straightforward matter to use (7.9), (8.1), (8.2) and (8.3) to obtain a classification. In order to distinguish the various cases it is helpful to use *Tits diagrams* for reductive algebraic groups over an arbitrary field. Indeed these diagrams also help to explain and illustrate Theorem (8.3) and Remarks 1 and 2 in section 3. Our discussion of these things will necessarily be rather sketchy because we shall avoid using algebraic groups!

First we consider quadratic forms. Let Δ be the building obtained using the totally singular subspaces of a quadratic form q of Witt index r on a vector space of dimension N over a commutative field K. When taken over a suitable extension L of K (for example if L is the algebraic closure of K) this form has Witt index $n = (N - 1)/2$ if N is odd, or $n = N/2$ if N is even. The corresponding building Δ_L is therefore of type $C_n(= B_n)$ if N is odd, and type D_n if N is even.

The building Δ is a subcomplex of Δ_L and its vertices belong only to the first r nodes of the Δ_L diagram; the Tits diagram for Δ is obtained by circling these nodes, so it is one of the following:

$$\text{(diagram)} \qquad B_n \qquad N = 2n+1$$

$$\text{(diagram)} \qquad D_n$$

$$\text{(diagram)} \qquad {}^2D_n \qquad \left.\right\} \quad N = 2n$$

The distinction between D_n and ${}^2D_n$ depends on the discriminant. If $r = 0$ the form is called *anisotropic* in which case Δ is vacuous, and there are no circled nodes. For example if $K = \mathbf{R}$, the dot product is anisotropic; furthermore for $K = \mathbf{R}$ and $N = 2n$ one has the D_n case when $n - r$ is even, and ${}^2D_n$ when $n - r$ is odd. We shall be particularly interested in

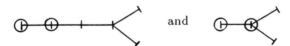

which represent quadratic forms of Witt index 2 on vector spaces of dimensions 12 and 8 respectively, over a commutative field.

Now let K be a non-commutative field having finite dimension over its centre k. If L is a maximal commutative subfield of K containing k, then $\dim_k L = \dim_L K = d$, so $\dim_k K = d^2$ (d is called the *degree* of K over its centre, and in fact L is a splitting field for K in the sense that $K \otimes_k L$ is a $d \times d$ matrix algebra over L). For example if K is the quaternions, then $k = \mathbf{R}$ and $L = \mathbf{C}$.

Consider the $A_r(K)$ building. Its vertices are subspaces of a vector space V of dimension $r + 1$ over K. They become certain subspaces of dimension $d, 2d, \ldots, rd$ when we take V as a $d(r + 1)$-dimensional vector space over L. Thus the $A_r(K)$ building is a subcomplex of the $A_{d(r+1)-1}(L)$

building, whose vertices have only those types circled in the following diagram:

We shall be particularly interested in the diagram  which represents a projective plane (A_2 building) over a quaternion algebra (i.e., $d = 2$). The subdiagram ⊢—⊕—⊣ represents a projective line over a quaternion algebra.

In general, given a pseudo-quadratic or sesquilinear form of Witt index r on a finite dimensional vector space over a field, where the field has finite degree d over its centre k, the diagram is B_n, C_n, D_n, ^2D_n or ^2A_n (see Appendix 2) in which nodes $d, 2d, \ldots, rd$ are circled. These are the cases which arise from algebraic groups, the finite-dimensional over k.

Non-Desarguesian Moufang Planes (Cayley Planes). As mentioned earlier, a non-Desarguesian, Moufang plane is coordinatised by a Cayley division algebra K, 8-dimensional over its centre k. We let K_0 denote a maximal commutative subfield of K; it has dimension 2 over k. For example if K is the Cayley numbers, then $k = \mathbf{R}$ and $K_0 = \mathbf{C}$. The points and lines of this Cayley plane can be taken as certain vertices in an $E_6(K_0)$ building (see [loc. cit.] (5.12), and earlier references cited there for more details); the two types of vertices are circled in the following diagram.

Each rank 1 residue is represented by a subdiagram of shape

In particular if L is a line, its points correspond to the totally singular subspaces of a quadratic form q of Witt index 1 on a 10-dimensional vector space W over k. If p is a point of L, and $\langle w \rangle$ the corresponding t.s. 1-space of W, then $\langle w \rangle^{\perp}/\langle w \rangle$ is an 8-space on which q is anisotropic, and the root group fixing all points of L and all lines on p is the additive group of this 8-space. This space also acquires a multiplicative structure (Appendix 1 section 2) making it a Cayley algebra, and the anistropic form induced by q is nothing other than the norm form of the Cayley algebra.

C_3 **Buildings having Cayley Planes.** We now explain Theorem 8.3 using diagrams. We have seen that a rank 1 residue of a Cayley plane has diagram

The only Moufang quadrangle having such a rank 1 residue is that with diagram

This quadrangle arises from a quadratic form q of Witt index 2 on a 12-dimensional vector space V, and if U is the 2-space corresponding to a line of the quadrangle, then q induces an anisotropic quadratic form on the 8-space $U^\perp/U$. After multiplication by a scalar, this is the norm form of the Cayley algebra for the plane. The diagram for the C_3 building is obtained by glueing the plane and quadrangle diagrams along their common residue to obtain the following form of E_7.

F_4 **Buildings.** Recall the F_4 diagram

$$\circ \underline{\quad} \circ = \!\!= \circ \underline{\quad} \circ$$

Both residues of the quadrangle are also residues of Moufang planes. If one of these is a Cayley plane, then the quadrangle is that given above, and the two C_3 subdiagrams are forced to be

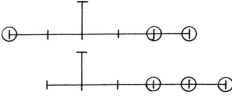

and

Identifying the common rank 2 (quadrangle) residue gives a form of E_8

This is the diagram for an F_4 building having a Cayley plane.

Now suppose both planes are Desarguesian. In this case the quadrangle must have the property that both types of vertices can be thought of as points, represented by totally singular (or isotropic) 1-spaces under a suitable form (see Remark 1 in section 3). This reduces us to four possible cases (see [loc. cit.] (10.10), and compare with the list of diagrams in Appendix 2). These are:

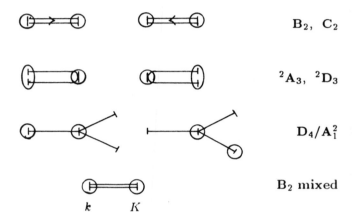

B_2, C_2

$^2A_3$, $^2D_3$

D_4/A_1^2

B_2 mixed

k K

The diagrams on the left are those arising from the quadratic form $n_K(x_o) - x_1 x_3 + x_2 x_4$ on $K \oplus k^4$ with $x_o \in K$, and $x_1, \ldots, x_4 \in k$, where k is commutative and K is one of:

(1) k itself, and $n_K(x_o) = x_0^2$ B_2

(2) a separable quadratic extension of k, with norm n_K $^2A_3$

(3) a quaternion algebra over k, with norm n_K D_4/A_1^2

Those on the right arise from a form f on a 4-dimensional vector space over K, where one of the following holds:

(1′) K is commutative, and f is alternating C_2

(2′) K is commutative, and f is $(\sigma, 1)$-hermitian with $[K : K^\sigma] = 2$ $^2D_3$

(3′) K is a quaternion algebra with centre k, and f is D_4/A_1^2
psuedo-quadratic (equivalently $(\sigma, -1)$-hermitian if
char $k \neq 2$) and $tr_{K/k}(x) = x + \sigma(x)$.

The final diagram represents the B_2 quadrangles of "mixed type" where $K \supset k \supset K^2$, and the quadratic form is $x_0^2 + x_1 x_3 + x_2 x_4$ where $x_0 \in K$ and $x_1, \ldots, x_4 \in k$.

Digression. We now briefly explain Remark 2 of section 3. In each quadrangle diagram of classical type, the line-residue has diagram

This represents a 2-dimensional vector space U over K, and the reversal of the diagram represents the dual vector space U^*. In every case except D_4/A_1^2 this subdiagram is connected at one end to the rest of the quadrangle diagram, and so this gives a preferred choice between U and U^*. However in the D_4/A_1^2 case

the residual diagram ⊢——⊕——⊣ has no preferred direction, so there is no distinction between U and its dual, and consequently no distinction between K and its opposite. In this special case K has degree 2 over its centre, or in other words is a quaternion algebra.

The F_4 Classification. Using the quadrangle diagrams above we are now able to write down the full classification of F_4 buildings, which the reader should check (Exercise 10). For completeness we include the diagram involving a Cayley plane, obtained earlier.

The full list of diagrams for F_4 buildings is

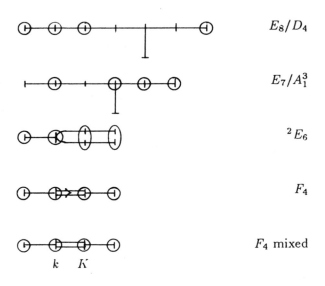

The last diagram is for char $k = 2$ and $K \supset k \supset K^2$; if $k = K$ this is the usual F_4 diagram.

6. Finite Buildings.

A finite building is always of spherical type because its apartments are finite Coxeter complexes. Assuming a connected diagram, the classification for rank ≥ 3 corresponds to the cases in sections 3-5 where the field is finite; and for rank 2, the classification of finite Moufang m-gons was done group theoretically by Fong and Seitz [1973] and [1974]. The groups concerned are called *finite groups of Lie type*, and are tabulated in Appendix 6 (this includes the rank 1 case for which the building is just a finite set of points). Our purpose here is to understand the effect of finiteness on the classification, and to remark on the order of the group and the fact that the subgroup U (of Chapter 6 section 4) is a Sylow-p-subgroup.

The fact that a finite field is commutative makes an immediate simplification; it implies for example that there is no F_4 building of type E_7/A_1^3 as this requires a quaternion algebra. Furthermore there is no finite Cayley division algebra, so this eliminates C_3 buildings having Cayley planes, and hence F_4 buildings of type E_8/D_4. The F_4 buildings of mixed type cannot occur either because finite fields are perfect; thus the F_4 classification reduces to two cases: $^2E_6(q)$ and $F_4(q)$, one for each ground field $\mathbf{F}_q$.

As to C_n buildings, the classification of non-degenerate (σ, ϵ)-hermitian forms on a vector space of dimension N over a finite field is well-known. If N is odd the Witt index is $(N - 1)/2$ and the form can be taken to be unitary ($\sigma \neq$ id.) or orthogonal ($\sigma =$ id.). If N is even, either the index is $\frac{N}{2}$ and the form is symplectic, unitary or orthogonal (group O_N^+), or the index is $\frac{N}{2} - 1$ and the form is orthogonal (group O_N^-). Given Witt index n and an arbitrary finite field $\mathbf{F}_q$ there is in each case a unique class of forms for which $\mathbf{F}_q$ is the fixed field of σ. Each of them gives a (thick) C_n building, except O_{2n}^+ which gives a D_n building. The finite simple groups are usually denoted respectively: $U_{2n+1}(q)$, $O_{2n+1}(q)$, $Sp_{2n}(q)$, $U_{2n}(q)$, $O_{2n}^+(q)$, $O_{2n+2}^-(q)$ - see Appendix 6.

To conclude this discussion we state a theorem.

(8.6) THEOREM. *The finite buildings having a connected diagram and rank ≥ 3 are those (of rank ≥ 3) listed in Appendix 6.*   □

Now let Δ be such a building (or a finite Moufang m-gon) and let p be the characteristic of the field. Let $G = \mathrm{Aut}\,\Delta$, and let B be the stabilizer of a chamber $c \in \Delta$. From Chapter 6 section 4, $B = UH$ where U is generated by a set of positive root groups and acts simple-transitively on the set of apartments containing c, and H is the pointwise stabilizer of such an apartment Σ. We shall demonstrate that U is a p-group and that $p \nmid |G : U|$. Thus U is a Sylow-p-subgroup of G, and B is a Sylow-p-normalizer (Exercise 16 of Chapter 6).

As a first step let $w = r_{i_1} \ldots r_{i_\ell}$ be a reduced expression for $w \in W$. Then the number of chambers d such that $\delta(c, d) = w$ is the same as the number of galleries of type $i_1 \ldots i_\ell$ starting at c, and this in turn equals the product $q_{i_1} \ldots q_{i_\ell}$ where $1 + q_{i_j}$ is the number of chambers in a residue of type i_j. In all but one case the chambers of such a residue correspond to the points of a projective line, or of a quadric, so q_i is a power of p (the exception is for $^2F_4$ where one type of residue is a Suzuki oval, but q_i is still a power of p). Thus the number of chambers d with $\delta(c, d) = w$ is a power of p. In particular this is true of chambers opposite p (equivalently apartments containing c), and so by (6.15) U is a p-group.

To continue our argument notice that for a panel π, any p-element fixing two chambers of $St(\pi)$ must fix a third (because $St(\pi)$ has $1 \pmod p$ chambers), and hence acts trivially on $St(\pi)$. Therefore by (6.4) any p-element fixing c and the apartment Σ acts trivially on Δ, so $p \nmid |H|$. Thus $p \nmid |B : U|$ and it remains to show that $p \nmid |G : B|$. In fact $|G : B|$ is the number of chambers and this is $1 \pmod p$ because, as shown above, the number of chambers at distance w from c is a power of p (greater than 1 if $w \neq 1$). We therefore have the following theorem.

(8.7) THEOREM. U is a Sylow-p-subgroup of the full automorphism group, and B is a Sylow-p-normalizer. $\qquad\square$

Remarks. Notice that if G is any group of automorphisms of Δ, containing U, the subgroup B stabilizing a chamber is a Sylow-p normalizer. As to H, it could contain p-elements, but only acting trivially on Δ of course. Notice also that we almost have a formula for the order of G. In the "untwisted" case where each rank 1 residue can be regarded as a projective line over the same field $\mathbf{F}_q$, the number of chambers opposite c is q^N where N is the length of the longest word. Furthermore the total number of chambers is $\sum_{w \in W} q^{\ell(w)}$. Thus

$$|G : H| = q^N \sum_{w \in W} q^{\ell(w)}$$

and the only imponderable is the order of H. If $G = GL_n(q)$ then H is the group of diagonal matrices, isomorphic to $\mathbf{F}_q^\times \times \ldots \times \mathbf{F}_q^\times$ (n copies); it contains the kernel of the action of G on Δ, namely the group of scalar matrices, isomorphic to $\mathbf{F}_q^\times$. In $PGL_n(q)$, H is the product of $n-1$ copies of $\mathbf{F}_q^\times$, one for each panel π of the chamber c (cf. Exercise 12 of Chapter 6). This is the usual form of H - for example in $E_8(q)$, H is isomorphic to a direct product of 8 copies of $\mathbf{F}_q^\times$. Finally we remark that the expression $\Sigma\, q^{\ell(w)}$ can be written as a product $\prod_{i=1}^{n} \frac{q^{d_i-1}}{q-1}$ where n is the rank. For this and for further details on these finite groups the standard reference is the book by Carter [1972].

Notes. The classification of spherical buildings (having a connected diagram of rank ≥ 3) is one of the principal objectives of Tits [1974], where the complete solution is given. The main difficulty concerned C_n buildings, described as polar spaces in Chapter 7, and classified in Chapters 8 and 9 of [loc. cit.]. The reduction to the C_3 case (8.4) was proved earlier by Veldkamp [1959] who determined all such polar spaces, except the ones involving Cayley planes, and those over non-commutative fields of characteristic 2, where the concept of a pseudo-quadratic form is needed; these forms were introduced by Tits [1974] Chapter 8. A very simple characterization of polar spaces is given by Buekenhout and Shult [1974]; the idea is that they are "point-line geometries" in which for each point p and line L, p is collinear with one or all points of L (though see their paper for other conditions on non-degeneracy and finite rank). The use of Tits diagrams for the classification of F_4 buildings appears in Chapter 10 of Tits [1974], and the diagrams themselves are introduced in Tits [1966]. Finally, in the case of single bond diagrams, Tits [1974] Chapter 5 uses the existence of algebraic groups of types E_6, E_7 and E_8 to obtain buildings of these types, and it was only recently (Ronan-Tits [1987]) that the buildings could be obtained independently (section 2 of this chapter). In fact Theorem (6.6) (proved in Chapter 4 of Tits [1974]) can now be used to obtain the groups from the buildings.

Exercises to Chapter 8

1. Given a $2n$-dimensional vector space with the quadratic form for a D_n building as in section 2, verify that every totally singular $(n-1)$-space is contained in exactly two t.s. n-spaces. [HINT: The orthogonal group

is transitive on t.s. $(n-1)$-spaces, by Witt's theorem - see e.g. Artin [1957] Theorem 3.9].

2. With the hypotheses of Exercise 1, let X, Y, X' be t.s. n-spaces such that $X \cap Y$ has dimension $n - 1$, and $X \cap X' = X \cap Y \cap X'$ has dimension $n - 2$. Show that $X' \cap Y$ has dimension $n - 1$.

3. Using Exercise 2, show that the graph whose vertices are t.s. n-spaces containing a fixed $(n-2)$-space, and whose edges are pairs X, Y where $\dim(X \cap Y) = n - 1$, is a complete bipartite graph.

4. Verify that the chamber system in section 2 obtained using the t.s. subspaces of dimension $\neq n - 1$ is indeed a chamber system of type D_n in the sense of Chapter 4. Show it is simply-connected, and hence a building. [HINT: For simple-connectivity use (4.10); any path is homotopic (in the topological sense) to one of points (1-spaces) and lines (2-spaces), and such paths are easily decomposed into triangles each of which lies in the residue of a 3-space].

5. With the notation of section 4, show that $K = K_{\sigma,\epsilon}$ if and only if $\sigma = $ id. and $\epsilon \neq 1$.

6. Let K be a field with centre k, and let q, f be as in section 4. Given an element $\lambda \in k$ with $\lambda + \lambda^\sigma = 1$, show that $q(x) = \lambda f(x, x) + K_{\sigma,\epsilon}$. Show such a λ exists if $\sigma|_k \neq$ id. [HINT: Expand $q(x(1 + \lambda))$ in two different ways].

7. If a sesquilinear form f arises from a pseudo-quadratic form q, show that f is trace-valued. In fact $f(x, x) = a + \epsilon a^\sigma$ where $a = q(x)$. [HINT: For $x \in V$, $t \in K$ expand $q(x(1 + t))$ in two different ways, and use $t^\sigma q(x) \equiv \epsilon q(x)^\sigma t \pmod{K_{\sigma,\epsilon}}$ to derive $(f(x, x) - q(x) - \epsilon q(x)^\sigma)t \in K_{\sigma,\epsilon}$ for all $t \in K$].

8. Observe that none of the exceptional Moufang quadrangles in Appendix 2 has a rank 1 residue which is the same as a rank 1 residue of a Moufang plane (and for this reason cannot form part of a C_3 building).

9. Consider the Tits diagram for a Desarguesian plane over a non-commutative field (i.e., $d \neq 1$). Show it is impossible to have three such diagrams sharing a common rank 1 diagram (this is a diagrammatic way of seeing that there is no D_4 building over a non-commutative field).

10. Verify that the list of diagrams for F_4 buildings is complete. In the E_7/A_1^3 case let k and K be the fields for the residual planes; what is the relationship between k and K?

Chapter 9
AFFINE BUILDINGS I

In this chapter we shall define affine buildings, and show that every affine building gives rise to a spherical building "at infinity". This building at infinity is a generalization of the "celestial" sphere at infinity of Euclidean space, whose points may either be taken as parallel classes of half-lines, or half-lines emanating from some fixed point.

1. Affine Coxeter Complexes and Sectors.

A building is called *affine* (or *of affine type*) if for each connected component of the diagram, the corresponding Coxeter complex can be realized as a triangulation of Euclidean space in which all chambers are isomorphic. Since any building is a direct product of buildings, one for each connected component of the diagram, nothing is lost by restricting attention to connected diagrams, and we shall do this. We remark however that for nonconnected diagrams, the Coxeter complex can be regarded as a tesselation of Euclidean space in which each chamber is a product of simplexes (e.g. in the $\tilde{A}_1 \times \tilde{A}_2$ case a chamber is a prism); in such cases the building can be described as a "polysimplicial complex" - Bruhat-Tits [1972].

The various classes of connected affine diagrams are listed below; the number of nodes is $n+1$ for type $\tilde{X}_n (X = A, \ldots , G)$, and the nodes marked by an s are the possible types of "special vertices" (explained later).

DIAGRAM   TYPE

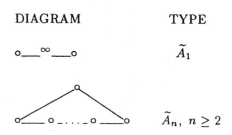

$$\tilde{A}_1$$

$$\tilde{A}_n, \ n \geq 2$$

DIAGRAM   TYPE

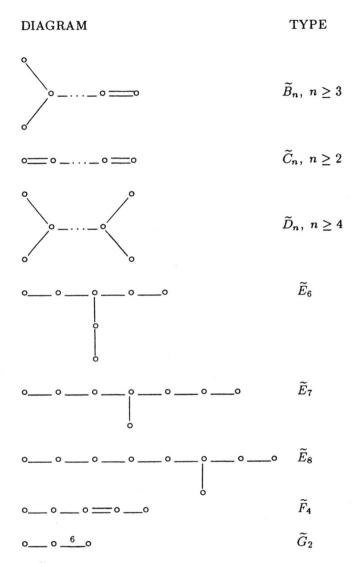

$\tilde{B}_n$, $n \geq 3$

$\tilde{C}_n$, $n \geq 2$

$\tilde{D}_n$, $n \geq 4$

$\tilde{E}_6$

$\tilde{E}_7$

$\tilde{E}_8$

$\tilde{F}_4$

$\tilde{G}_2$

In the $\tilde{A}_1$ case the Coxeter complex is nothing other than a doubly infinite sequence of chambers $\cdots c_{-1}, c_0, c_1, c_2, \cdots$ each of which is adjacent to its two neighbors, and this can obviously be realized as the Real line with integer points as panels and unit intervals as chambers. For the other diagrams, which have at least three nodes, each chamber can be taken as a Euclidean simplex such that for each $i, j \in I$ the angle between the i-face and the j-face is π/m_{ij}. For example, if $I = \{1, 2, 3\}$ then since the sum of the angles of a Euclidean triangle is π, one has:

$$\frac{1}{m_{12}} + \frac{1}{m_{13}} + \frac{1}{m_{23}} = 1,$$

giving the diagrams $\tilde{A}_2$, $\tilde{C}_2$ and $\tilde{G}_2$.

More generally consider an n-dimensional Euclidean simplex c in $\mathbf{R}^n$, whose codimension 1 faces are labelled by elements of an $(n + 1)$-set I, such that the angle between the i-face and j-face is π/m_{ij} for some integer m_{ij}. The group generated by reflections in the codimension 1 faces of c is the Coxeter group W of type $M = (m_{ij})$. Indeed if s_i is the reflection in the i-face, then $s_i s_j$ has order m_{ij} and hence the s_i certainly generate a quotient of W. This shows that the Coxeter complex maps onto $\mathbf{R}^n$, and once this map is shown to be a homeomorphism in the neighborhood of each point (see Exercise 6), it follows from the simple-connectivity of $\mathbf{R}^n$ that the map is an isomorphism of simplicial complexes, and the s_i generate W itself. This also shows that a connected diagram is affine precisely when there exists a Euclidean simplex whose dihedral angles are π/m_{ij}.

In its action on $\mathbf{R}^n$, W is a discrete subgroup of the group of affine isometries, of shape $\mathbf{R}^n \cdot O(n)$, where $\mathbf{R}^n$ is the normal subgroup of translations, and $O(n)$ is the orthogonal group, stabilizing a point. Thus W has a normal subgroup $\mathbf{Z}^n$ of translations whose quotient W_o, being a discrete subgroup of the compact group $O(n)$, is finite. Moreover since W is generated by reflections, W_o is generated by images of these reflections; but a finite group generated by reflections is a Coxeter group and in this case it is generated by just n linearly independent reflections (see Bourbaki [1968/81] Ch. V, section 3.9, Prop. 7, p.85). Let $s_1, \ldots, s_n$ be reflections in W, whose images in W_o generate W_o, and let $M_1, \ldots, M_n$ be the walls fixed by $s_1, \ldots, s_n$ respectively. Since the M_i have codimension 1 and are linearly independent their intersection is a vertex. Such vertices are called *special*.

Notice that any finite subgroup of W maps (via $W \to W/\mathbf{Z}^n$) isomorphically into W_o, because $\mathbf{Z}^n$ contains no non-identity elements of finite order. Therefore using the orders of finite Coxeter groups in Appendix 5, it is a simple matter to check which vertices are special.

Sectors. Let s denote a special vertex, and c a chamber having s as one of its vertices. The panels of c having s as a vertex determine roots $\alpha_1, \ldots, \alpha_n$ containing c, and their intersection $S = \alpha_1 \cap \alpha_2 \cap \ldots \cap \alpha_n$ is called a *sector*

(French: *quartier*) with *vertex s* and *base chamber c*. In terms of the affine space structure, a sector is a simplicial cone; Figure 9.1 shows an example in the $\tilde{G}_2$ case.

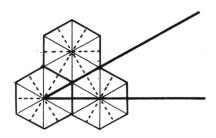

Figure 9.1

That part of a wall bounding a sector (e.g., $\partial \alpha_1 \cap \alpha_2 \cap \ldots \cap \alpha_n$) will be called a *sector-panel* (French: *cloison de quartier*).

(9.1) LEMMA. *If the sector S, having vertex s, contains the sector T, then S is the convex hull of s and T.*

PROOF: Let V denote the convex hull of s and T, which by (2.8) is an intersection of roots. Let α be any root containing s and T, and let $\alpha' \subset \alpha$ be a root having s on its boundary. Since the boundary walls $\partial \alpha$ and $\partial \alpha'$ must be parallel, the strip $\alpha - \alpha' = \alpha \cap (-\alpha')$ cannot contain a sector. In particular $T \not\subset -\alpha'$, hence $S \not\subset -\alpha'$, and therefore S lies in α' (it must lie in α' or $-\alpha'$, since $s \in \partial \alpha'$). Therefore V lies in α', and is hence an intersection of roots whose boundary walls contain s. Thus V is a simplicial cone, and since S is a minimal simplicial cone (having only one base chamber) we conclude that $V = S$. ☐

(9.2) LEMMA. *Given sectors S and S' in an affine Coxeter complex, S' is a translate of S if and only if $S \cap S'$ contains a sector, in which case $S \cap S'$ is a sector. In particular if S contains subsectors S_1 and S_2, then $S_1 \cap S_2$ is a sector.*

PROOF: If S' is a translate of S, then it is a straightforward exercise to show that $S \cap S'$ is a sector (Exercise 2). To prove the converse it suffices to show that if a sector S contains a sector T, then S is a translate of T.

Indeed let g be a translation for which $g(T)$ has the same vertex as S. Then $g(T) \cap T$ is a sector lying in both $g(T)$ and S, and since these have the same vertex, (9.1) implies $g(T) = S$. To prove the final statement notice that S_1 and S_2 are both translates of S, hence S_2 is a translate of S_1, so $S_1 \cap S_2$ is a sector.   □

2. The Affine Building $\widetilde{A}_{n-1}(K, v)$.

The Discrete Valuation. Let K be a field (not necessarily commutative) with a *discrete valuation* v; after multiplying v by a suitable positive real number this means that we have a surjective map $v : K^\times \to \mathbf{Z}$ satisfying

$$v(ab) = v(a) + v(b)$$
$$v(a + b) \geq \min(v(a), v(b))$$

for all $a, b \in K$, with the convention that $v(0) = +\infty$. Let $\mathcal{O}$ denote the *valuation ring* of K with respect to v:

$$\mathcal{O} = \{a \in K | v(a) \geq 0\}.$$

This ring has a unique maximal ideal m:

$$m = \{a \in K | v(a) \geq 1\}.$$

We let $\pi \in K^\times$ be a *uniformiser*, i.e., $v(\pi) = 1$. For each $a \in K^\times$ one has

$$a\mathcal{O} = \mathcal{O}a = \pi^{v(a)}\mathcal{O} = \{x \in K | v(x) \geq v(a)\}.$$

In particular the ideals of $\mathcal{O}$ are the m^ℓ where $\ell = 1, 2, \ldots$. We let k denote the *residue field* $\mathcal{O}/m = \mathcal{O}/\pi\mathcal{O}$. For details on fields having a discrete valuation, see for example the book *Local Fields* by Serre [1962/79].

Exercise. If $v(a) < v(b)$, show that $v(a + b) = v(a)$.

Example 1. Let $K = \mathbf{Q}$ (the rational numbers), and let p be a prime. Every rational can be written as $p^n a/b$ where a and b are integers not divisible by p. We set $v(p^n a/b) = n$. The valuation ring is the ring $\mathbf{Z}_{(p)}$ of integers localised at the prime ideal (p), and the residue field is the finite field $\mathbf{Z}/(p)$ of integers modulo p. Every discrete valuation of $\mathbf{Q}$ is obtained in this way for some prime p.

Example 2. Let k be a commutative field and let $K = k(t)$, the field of rational functions in one variable. If f and g are polynomials then $v_\infty\left(\frac{f}{g}\right) = \deg g - \deg f$ is a discrete valuation. Moreover one obtains discrete valuations v_a for each element $a \in k$, as follows. Any rational function can be written $(t-a)^n \frac{f}{g}$ where f and g are polynomials not divisible by $(t-a)$; set $v_a\left((t-a)^n \frac{f}{g}\right) = n$. If k is algebraically closed, every discrete valuation which is trivial on k is v_∞ or v_a for some $a \in k$.

Lattices. Let V be an n-dimensional vector space over K. A v-lattice (or simply a lattice) of V will mean any finitely generated $\mathcal{O}$-submodule of V which generates the K-vector space V; such a module is free of rank n. If L is a lattice, and $a \in K^\times$, then since $a\mathcal{O} = \mathcal{O}a$, we see that aL is also a lattice. We shall call two lattices *equivalent* if one is a multiple of the other in this way; this is clearly an equivalence relation, and we let [L] denote the equivalence class of the lattice L.

We now define the building Δ as a simplicial complex as follows. Its vertices are the classes [L] of lattices. Its edges are the unordered pairs of vertices x and y such that if L is in the class of x, there is an L' in the class of y for which $\pi L \subset L' \subset L$; the existence of such an L' is independent of the lattice L chosen to represent x. The simplexes are given by sets of vertices any two of which lie on a common edge. Such sets of vertices can be written as $[L_1], [L_2], \ldots, [L_t]$ where

$$L_1 \supset L_2 \supset \ldots \supset L_t \supset \pi L_1. \tag{$\dagger$}$$

Since $L_i/\pi L_1$ is a subspace of the n-dimensional k-vector space $L_1/\pi L_1$ (k is the residue field $\mathcal{O}/\pi\mathcal{O}$), one sees that maximal simplexes have n vertices - these are the *chambers*.

We shall show (below) that there are n different types of vertices, and chambers have one of each. We therefore define two chambers to be *i-adjacent* if they differ in at most a vertex of type i. By ($\dagger$) above the residue of a vertex ([L_1] in that case) is isomorphic to the $A_{n-1}(k)$ building, whose simplexes are the flags of an n-dimensional k-vector space ($L_1/\pi L_1$ in that case). Therefore Δ is a chamber system of type $\tilde{A}_{n-1}$, and by (4.10) it will be a building once it is shown to be simply-connected as a simplicial complex. This is done in Exercise 10, using apartments. A thorough discussion of the $n = 2$ case (where Δ is a tree) is given in Serre's book on *Trees* [1977/80] Chapter II section 1.

Some Subgroups of $GL(V)$. The group $GL(V)$ (i.e., $GL_2(K)$) acts transitively on the set of all $\mathcal{O}$-lattices, and preserves equivalence between lattices. Its subgroup $SL(V)$ can be defined in one of the following equivalent ways (see e.g., Artin [1957] Chapter 4).

(i) It is the subgroup of $GL(V)$ generated by root groups (unipotent elements).

(ii) It is the commutator subgroup $[GL(V), GL(V)]$.

(iii) It is the kernel of the *Dieudonné determinant*

$$\det : GL(V) \to K^\times/[K^\times, K^\times].$$

Since the valuation v is trivial on $[K^\times, K^\times]$, one has for each $g \in GL(V)$ a well-defined integer $v(\det(g))$. We shall write:

$$GL(V)^o = \{g \in GL(V) | v(\det(g)) = 0\}$$
$$GL(V)^{o(n)} = \{g \in GL(V) | v(\det(g)) \equiv 0 (\mathrm{mod}\ n)\}.$$

Obviously:

$$SL(V) \subset GL(V)^o \subset GL(V)^{o(n)} \subset GL(V).$$

Consider now the stabilizer of a vertex $x = [L]$. If $g \cdot x = g$, then $gL = cL$ for some $c \in K^\times$, so $v(\det(g)) = n \cdot v(c) \equiv 0 (\mathrm{mod}\ n)$. If $g \in GL(V)^o$, then $v(c) = 0$, so c is a unit in $\mathcal{O}$, and $L = cL$. Thus, using G_a to denote the stabilizer of a, we have:

$$\text{If } G \text{ is a subgroup of } GL(V)^o, \text{ then } G_x = G_L \qquad (*)$$

Types. We define the *type* of a vertex as an integer mod n. Start with some lattice L and assign type 0 to $[L]$. If L' is any lattice, then $L' = gL$ for some $g \in GL(V)$, and we define $[L']$ to have *type* $v(\det(g)) \mathrm{mod}\ n$. By the discussion above this is well-defined mod n, and $GL(V)^{o(n)}$ preserves types.

Consider a chamber, represented by $L_1 \supset \ldots \supset L_n \supset \pi L_1$; regarding $L_i/\pi L_1$ as a subspace of $L_1/\pi L_1$, we immediately find a basis $e_1, \ldots, e_n$ for V such that:

$$L_1 = \langle e_1, \ldots e_n \rangle_{\mathcal{O}}$$
$$L_2 = \langle \pi e_1, e_2, \ldots e_n \rangle_{\mathcal{O}}$$
$$\vdots$$
$$L_n = \langle \pi e_1, \ldots, \pi e_{n-1}, e_n \rangle_{\mathcal{O}}$$

If $gL_1 = L_i$ then $v(\det(g)) = i - 1$, so the n vertices of a chamber have n different types. As explained above this allows us to regard Δ as a chamber system.

Exercise. Show that $SL(V)$ is transitive on the set of vertices of a given type.

Bounded Subgroups. Given a basis for V, every element $g \in GL(V)$ can be written as a non-singular $n \times n$ matrix (g_{ij}). A subgroup G of $GL(V)$ will be called *bounded* if there is an integer d such that $v(g_{ij}) \geq d$ for all $(g_{ij}) = g \in G$ (obviously $d \leq 0$ because $v(1) = 0$). This definition is independent of the choice of basis (though d itself is, of course, not independent of the basis). The $\geq$ sign is due to the fact that each element $c \in K$ has a "norm" $|c| = e^{-v(c)}$, so $v(c)$ is bounded below if and only if $|c|$ is bounded above.

(9.3) THEOREM. *A subgroup G of $GL(V)^\circ$ is bounded if and only if it stabilizes a vertex of Δ. Furthermore the vertices of Δ are in bijective correspondence with the maximal bounded subgroups of $GL(V)^\circ$.*

PROOF: If $x = [L]$ is any vertex, take an $\mathcal{O}$-basis for L; using this as a basis for V, we have $G_L \leq GL_n(\mathcal{O})$, which is bounded (using $d = 0$). Thus by (*) above $G_x = G_L$ is bounded. Conversely if G is a bounded subgroup of $GL(V)^\circ$, take any lattice L and set

$$L_0 = \sum_{g \in G} gL.$$

Since G is bounded, L_0 is also a lattice, and is stabilized by G.

For the final statement of the theorem it suffices to show that if $G = GL(V)^\circ$, then G_L fixes no vertices apart from $x = [L]$. Since $G_L \cong GL_n(\mathcal{O})$ acts as $GL_n(k)$ on $St(x)$ (recall k is the residue field $\mathcal{O}/\pi\mathcal{O}$), it fixes nothing in $St(x)$, and hence by Exercise 12 fixes no other vertex. $\qquad\square$

Apartments. Take a basis $e_1, \ldots, e_n$ of V, and let A be the subcomplex of Δ whose vertices are all $[L]$, where $L = \langle \pi^{r_1} e_1, \ldots, \pi^{r_n} e_n \rangle_\mathcal{O}$ is the $\mathcal{O}$-lattice spanned by $\pi^{r_1} e_1, \ldots, \pi^{r_n} e_n$. Without loss of generality we may scale $e_1, \ldots, e_n$ so that $\langle e_1, \ldots, e_n \rangle_\mathcal{O}$ has type 0, in which case L has type $r (\bmod n)$, where $r = r_1 + \ldots + r_n$.

Notice that $[L]$ is equivalent to the set of n-tuples $(r_1 + t, \ldots, r_n + t)$ for $t \in \mathbf{Z}$, and hence is equivalent to the single n-tuple $(x_1, \ldots, x_n)$ where $x_i = r_i - \frac{r}{n}$. Thus the vertices of A correspond to certain points of $\mathbf{R}^n$

lying in the hyperplane $x_1 + \ldots + x_n = 0$. To see that these are the vertices of the Coxeter complex of type $\widetilde{A}_{n-1}$, it suffices to check that the following involutions $s_1, \ldots, s_n$ preserve this structure and satisfy the relations required by a Coxeter group of type $\widetilde{A}_{n-1}$: s_i switches x_i with x_{i+1} if $i = 1, \ldots, n-1$, and s_n replaces x_n by $x_1 + 1$, and x_1 by $x_n - 1$.

Exercise. After a suitable rescaling and reordering of the basis $e_1, \ldots, e_n$ show that the vertices in a sector are those $[L]$ for which $L = \langle e_1, \pi^{r_2} e_2, \ldots, \pi^{r_n} e_n \rangle_{\mathcal{O}}$ where $0 \leq r_2 \leq \ldots \leq r_n$.

The Affine Tits System. After choosing a suitable basis for V, a chamber stabilizer B in $GL(V)^{o(n)}$ is the inverse image of the group of upper triangular matrices under the projection from $\mathcal{O}$ to $\mathcal{O}/\pi\mathcal{O} = k$. Thus

$$B = \begin{bmatrix} \mathcal{O} & & \mathcal{O} \\ & \ddots & \\ \pi\mathcal{O} & & \mathcal{O} \end{bmatrix}$$

The stabilizer N of an apartment is almost the same as that for the spherical building $A_{n-1}(K)$, namely permutation matrices times diagonal matrices, except that we must ensure N is a subgroup of $GL(V)^{o(n)}$ otherwise it will not preserve types. The panel stabilizers (minimal parabolics) are:

$$P_0 = \begin{bmatrix} \mathcal{O} & & \pi^{-1}\mathcal{O} \\ & \ddots & \\ \pi\mathcal{O} & & \mathcal{O} \end{bmatrix}$$

$$P_i = \begin{bmatrix} \mathcal{O} & & & & \mathcal{O} \\ & \ddots & & & \\ & & \boxed{\begin{matrix} \mathcal{O} & \mathcal{O} \\ \mathcal{O} & \mathcal{O} \end{matrix}} & & \\ & & & \ddots & \\ \pi\mathcal{O} & & & & \mathcal{O} \end{bmatrix}$$

where P_0 differs from B only in the $(1, n)$ entry, and for $i = 1, \ldots, n-1$, P_i differs only in the $(i+1, i)$ entry, and has a $GL_2(\mathcal{O})$ block on the diagonal as shown. Obviously $\langle P_1, \ldots, P_{n-1} \rangle = GL_n(\mathcal{O})$ stabilizes a vertex.

Root Groups in an Affine Moufang Building. In Chapter 6 section 4 we introduced Moufang buildings, and we can now give an example of affine type. Suppose K contains its residue field k as a subfield; e.g., $K = k(t)$ as in Example 2, with valuation $v = v_0$ determined by $v(t) = 1$. Consider the $\widetilde{A}_2(K, v)$ building, and take "root groups" of the form

$$U_\alpha = \begin{bmatrix} 1 & at^r & 0 \\ 0 & 1 & 0 \\ 0 & 0 & 1 \end{bmatrix}$$

where r is fixed, and a ranges over the residue field k; obviously U_α is isomorphic to the additive group of k. The root groups $U_i = U_{\alpha_i}$ for the "fundamental roots" α_i (see Chapter 7 section 2) are the following, where a, b, c range over the residue field k:

$$U_1 = \begin{bmatrix} 1 & a & 0 \\ 0 & 1 & 0 \\ 0 & 0 & 1 \end{bmatrix} \qquad U_{-1} = \begin{bmatrix} 1 & 0 & 0 \\ a & 1 & 0 \\ 0 & 0 & 1 \end{bmatrix}$$

$$U_2 = \begin{bmatrix} 1 & 0 & 0 \\ 0 & 1 & b \\ 0 & 0 & 1 \end{bmatrix} \qquad U_{-2} = \begin{bmatrix} 1 & 0 & 0 \\ 0 & 1 & 0 \\ 0 & b & 1 \end{bmatrix}$$

$$U_3 = \begin{bmatrix} 1 & 0 & 0 \\ 0 & 1 & 0 \\ ct & 0 & 1 \end{bmatrix} \qquad U_{-3} = \begin{bmatrix} 1 & 0 & ct^{-1} \\ 0 & 1 & 0 \\ 0 & 0 & 1 \end{bmatrix}$$

We may let B be the same as before, but must restrict N to permutation matrices times diagonal matrices of the form

$$\begin{pmatrix} at^{r_1} & 0 & 0 \\ 0 & bt^{r_2} & 0 \\ 0 & 0 & ct^{r_3} \end{pmatrix}$$

where $a, b, c \in k^\times$, and of course $r_1 + r_2 + r_3 \equiv 0 (\mathrm{mod}\ 3)$ in order that N be type preserving. It is left to the reader to check that the U_α and N satisfy (M1) - (M4) of Chapter 6 section 4. Before leaving this example, notice that the U_α are not unique. We could equally well have chosen some other rational function f, with $v(f) = 1$, in place of t.

Exercise. Given $e_i \in U_i - \{1\}$, and $n_i = m(e_i) \in U_{-i} e_i U_{-i} \cap N$, in this example, show that $n_1 n_2 n_1 = n_2 n_1 n_2$, $n_1 n_3 n_1 = n_3 n_1 n_3$, and $n_2 n_3 n_2 = n_3 n_2 n_3$ (cf. (7.3) and Appendix 1 (A.5)).

Completion. Let $\widehat{K}$ be the completion of K with respect to v; it is the quotient field of its valuation ring $\widehat{\mathcal{O}} = \varprojlim \mathcal{O}/\pi^n\mathcal{O}$. (Completing K to $\widehat{K}$ is the same as adjoining all limits of Cauchy sequences, where the distance between two elements x and y is $|x-y| = e^{-v(x-y)}$ - in particular $|x-y| \to 0$ when $v(x-y) \to \infty$). For example if $K = \mathbf{Q}$ with the p-adic valuation, as in Example 1 above, then $\widehat{K} \cong \mathbf{Q}_p$ with valuation ring $\mathbf{Z}_p$ (the p-adic integers); if K is a function field of degree 1, for example $k(t)$ as in Example 2, then $\widehat{K} \cong k((t))$ with valuation ring $k[[t]]$ (the ring of formal power series).

We set $\widehat{V} = V \otimes_K \widehat{K}$, and associate to each lattice L of V the lattice $\widehat{L} = L \otimes_{\mathcal{O}} \widehat{\mathcal{O}}$ of $\widehat{V}$. This gives a bijection of the set of $\mathcal{O}$-lattices of V onto the set of $\widehat{\mathcal{O}}$-lattices of $\widehat{V}$, showing that the affine building Δ of V is isomorphic to that of $\widehat{V}$. This fact can also be seen geometrically because the building obtained from V obviously embeds in that obtained from $\widehat{V}$, yet their residues are isomorphic (because the residue field is the same in both cases), so the embedding is an equality.

There is however an important difference which will be made precise in the next chapter when we deal with apartment systems. Each basis of $\widehat{V}$ gives an apartment of Δ, as explained earlier, and in fact all apartments of Δ (i.e., isometric images of the $\widetilde{A}_{n-1}$ Coxeter complex) arise in this way - see Exercise 2 of Chapter 10. The bases of V do not give all possible apartments, only those in a particular "apartment system" $\mathcal{A}$. Using all apartments of Δ we shall, in the next section, obtain a "building at infinity" Δ^∞, isomorphic to the spherical building $A_{n-1}(\widehat{K})$. Using only those apartments in $\mathcal{A}$, one obtains a smaller "building at infinity" $(\Delta, \mathcal{A})^\infty$, isomorphic to the spherical building $A_{n-1}(K)$.

3. The Spherical Building at Infinity.

A sector has been defined in an affine Coxeter complex or apartment; we now define a *sector* of an affine building to mean a sector in some apartment of the building. Of course if S is a sector in some apartment then it is a sector in any apartment containing it, since the two apartments are isometric via an isometry fixing their intersection (Exercise 6 of Chapter 3).

(9.4) LEMMA. *Given any chamber c, and any sector S, there exists a sector $S_1 \subset S$ such that S_1 and c lie in a common apartment.*

PROOF: Let A be an apartment containing S, and assume $c \notin A$. By induction along a gallery from c to A it suffices to prove the lemma when

c has a panel π in A. Of the two roots of A whose boundary contains π, let α be the one containing a sector $S_1 \subset S$ (see Figure 9.2).

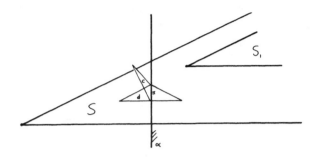

Figure 9.2

If d is the chamber of $A - \alpha$ on π, then clearly $\{c\} \cup \alpha$ is isometric to $\{d\} \cup \alpha$, and hence by (3.6) c and α lie in a common apartment. $\square$

The retraction $\rho_{S,A}$. Now let c, S and S_1 be as in the preceding lemma, and let A be any apartment containing S. The fact that $\{c\} \cup S_1$ is isometric to a subset of W, shows that for any chambers $x, y \in S_1$, one has $\rho_{x,A}(c) = \rho_{y,A}(c)$. We let $\rho_{S,A}(c)$ denote this common chamber of A; it is independent of S_1 because any two subsectors of S intersect non-trivially. (If we treat Δ as a simplicial complex then $\rho_{S,A}$ is a retraction of Δ onto A).

(9.5) PROPOSITION. *Any two sectors S and T contain subsectors S_1 and T_1 lying in a common apartment.*

PROOF: The proof is deferred to section 4. $\square$

We now define a *sector-face* to mean a face of a sector treated as a simplicial cone; thus sector-faces are themselves simplicial cones, and those of codimension 1 are the sector-panels. Two sector-faces, or walls, are said to be *parallel* if the distance between them is bounded (i.e., if the distance from any point of one to the nearest point of the other is bounded).

Obviously parallelism is an equivalence relation, and in a given apartment two walls, or sector-faces, are parallel if one is a translate of the other. As a matter of notation we let X^∞ denote the parallel class of X, and sometimes call it the *direction* of X, or the *trace* of X *at infinity*.

Using (9.2) and (9.5) it is straightforward to see that two sectors are parallel if and only if their intersection contains a sector (Exercise 3). In

Figure 9.3 the sectors S and T are parallel and intersect in the cross-shaded area; the sector-panels p_1 and q_1 of S and T are parallel, as are p_2 and q_2. In this example p_2 and q_2 contain a sector-panel in common, but p_1 and q_1 do not; this distinction will be important in Chapter 10 when we define two sector-panels to be asymptotic if their intersection contains a sector-panel (a refinement of parallelism).

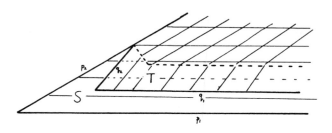

Figure 9.3

We now define the building at infinity, Δ^∞, as a chamber system over I_o where $I_o = I - \{o\}$, o being some fixed type of special vertices. The chambers of Δ^∞ are defined to be parallel classes of sectors of Δ, and two chambers c and d are *adjacent* if there are representative sectors S and T (i.e., $c = S^\infty$, $d = T^\infty$) having sector-panels D and E which are parallel; this is independent of the choice of S and T, because if $c = (S')^\infty$ then S' has a sector-panel parallel to D, and hence to E. Evidently the panels of Δ^∞ are parallel classes of sector-panels, and we determine the type $i \in I_o$ of a panel as follows. In each parallel class take a sector-panel having a vertex of type o; its base panel, the one on the vertex, must have some type $i \in I_o$: we take this to be the *type* of the parallel class. To check that this is well-defined it suffices, by (9.5), to check it in a single apartment, so consider two parallel sector-panels in a common apartment. They are translates of one another, and if they both have vertices of the same type (o in our case), then the translation may be done by an element of W, and hence their base panels have the same type.

To show Δ^∞ is a building we use apartments as in (3.11). Given an apartment A of Δ, let $A^\infty = \{S^\infty | S$ a sector in $A\}$. Then A^∞ is a Coxeter complex of type W_o because if s is a vertex of A of type o, every parallel class of sectors in A contains a unique sector having vertex s, and so A^∞ is isomorphic to $St(s)$. By (9.5) any two chambers of Δ^∞ lie in a common A^∞. Moreover let x be a chamber and y a chamber or panel of Δ^∞ contained in A_1^∞ and A_2^∞; then again by (9.5) $x = X^\infty$ and $y = Y^\infty$ where X and Y lie in $A_1 \cap A_2$. Now by Exercise 6 of Chapter 3 there is an isomorphism from A_1 to A_2 fixing $A_1 \cap A_2$, and hence an isomorphism from A_1^∞ to A_2^∞ fixing x and y. Thus by (3.11) Δ^∞ is a building.

(9.6) THEOREM. *With the above notation Δ^∞ is a building of spherical type M_o. Moreover the apartments of Δ^∞ are in bijective correspondence with the apartments of Δ, via $A \to A^\infty$. The faces of Δ^∞, considered as a simplicial complex, are the parallel classes of faces of Δ, and the walls of Δ^∞ are parallel classes of walls of Δ.*

PROOF: The first statement has already been proved. To prove the second, let X be an apartment of the spherical building Δ^∞, and let c and d be opposite chambers of X. Then by (9.5) we can find sectors S and T in the directions c and d, and lying in a common apartment A. Thus c and d lie in A^∞, but being opposite they lie in a unique apartment (see (2.15)(iv) or Exercise 5 of Chapter 3), so $A^\infty = X$. Moreover if $(A')^\infty = X$ then after possibly replacing S and T by subsectors we may assume they lie in A' also. Since $A \cap A'$ is convex (see (3.8)), either $A = A'$ or $A \cap A'$ is an intersection of half-apartments, by (2.8); but S and T do not lie in a common half-apartment, so $A = A'$. The third statement is left as an exercise.   □

Having defined Δ^∞ by using parallel classes of sectors, and their faces, we shall now show that it could equally well be obtained by choosing a special vertex s and using those sectors, and their faces, having vertex s.

(9.7) LEMMA. *If s is a special vertex, then each parallel class of sectors contains a unique sector having vertex s.*

PROOF: If S is any sector in the given class, then by (9.4) it contains a subsector S_1 lying in a common apartment A with s. The translate of S_1 in A having vertex s is then parallel to S by (9.2). It remains to show that if two parallel sectors S and T have the same vertex s, then they are equal. This follows from (9.1) because S and T are both equal to the convex hull of s and some sector R contained in $S \cap T$.   □

Let us now suppose that our special vertex s has type o and, as before, define a sector-panel with vertex s to have *type i* if its base panel (the one on s) has type i. Two sectors having vertex s will be called i-*adjacent* if they share such a sector-panel.

(9.8) THEOREM. *The sectors having vertex s, together with the adjacencies just defined, form a chamber system isomorphic to Δ^∞.*

PROOF: Obviously if two sectors S and T having vertex s are i-adjacent then S^∞ and T^∞ are i-adjacent in Δ^∞. Therefore in view of (9.7) it suffices to prove that for sectors S and T having vertex s, if S^∞ and T^∞ are i-adjacent in Δ^∞, then S is i-adjacent to T.

Let A_1 be any apartment containing a subsector $S_1 \subset S$; if α is a half-apartment of A_1, minimal with respect to containing S_1, then α and s lie in a common apartment. Now by (9.5) choose A_1 to contain subsectors $S_1 \subset S$ and $T_1 \subset T$, and then take α to be a root containing them both (this is possible since S_1^∞ and T_1^∞ are adjacent in Δ^∞). Hence there is an apartment A containing S_1, T_1 and s. By (9.1) S is the convex hull of s and S_1, and T is the convex hull of s and T_1, so S and T are sectors in A having a common vertex. Since S^∞ and T^∞ are adjacent in Δ^∞ they have sector-panels which are parallel, but these sector-panels lie in A and have the same vertex; hence they are equal, and S is adjacent to T.   □

4. The Proof of (9.5).

To prove (9.5) we use "sector directions" which we now define. Let c be a chamber in an apartment A, which we treat as Euclidean space. If S is any sector of A, and s its vertex, take an ϵ-neighborhood of s in S, translate it to the barycentre of c, and call it $S(c)$. Here ϵ should be small enough so that a ball of radius ϵ lies entirely inside c. We call $S(c)$ the *sector direction* of S at c. It is independent of A, because if A' is any other apartment containing c and S, then $\rho_{c,A}$ maps A' to A, preserving the Euclidean space structure and fixing c and S. Notice that the set of sector directions at c is in bijective correspondence with the set of chambers on a special vertex, and hence corresponds to the elements of the finite Coxeter group W_o. If M is a wall dividing the apartment A into two roots $\pm\alpha$, then a sector direction will be said to be *on the $+\alpha$* (or $-\alpha$) *side* of M if, after translating its vertex to a point of M, it lies in $+\alpha$ (or $-\alpha$).

Before proving (9.5) we obtain a subsidiary result.

(9.9) LEMMA. *If S is a sector in an apartment A, and if T is any sector, then T contains a subsector T_1 such that $\rho_{S,A}|T_1$ is an isometry.*

PROOF: Write $\rho = \rho_{S,A}$. Since ρ preserves adjacency of chambers it is a question of finding a subsector T_1 such that for any two distinct and adjacent chambers $x, y \in T_1$, $\rho(x) \neq \rho(y)$. Indeed since T_1 is convex, any two chambers $z, z' \in T_1$ are joined by a gallery in T_1 of reduced type f, and ρ sends this to another gallery of type f, hence $\delta(\rho(z), \rho(z')) = \delta(z, z')$. Now let π be the panel common to x and y, and take x to be nearer the base chamber of T. We set $x' = \rho(x)$, $y' = \rho(y)$, $\pi' = \rho(\pi)$, and let α be the root of A containing x' but not y' - see Figure 9.4.

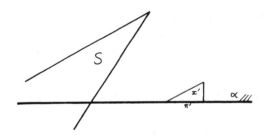

Figure 9.4

For any chamber $z \in T$, consider the two sector directions

$$S(\rho(z)) \text{ and } \rho(T(z)).$$

Without loss of generality let $S(\rho(z))$ correspond to $1 \in W_o$, and let $w(z) \in W_o$ denote $\rho(T(z))$.

Step 1. If $x' \neq y'$ then $w(x) = w(y)$.

Indeed if $x' \neq y'$ then ρ restricted to $\{x, y\}$ is an isomorphism.

Step 2. If $x' = y'$ then $w(y) = rw(x)$ where r is the reflection of W_o determined by the wall M of A.

This follows from Step 1 since $\rho|\{x, y\}$ may be taken as an isomorphism followed by a folding across M.

Step 3. If $x' = y'$ then S contains no subsector in α.

Suppose S contains a subsector in α. By (9.2) any two subsectors of S intersect non-trivially, so there is a chamber $c \in \alpha \cap S$ such that $\rho_{c,A}$

agrees with ρ on $\{x,y\}$. Since $c \in \alpha$, we have $x' = y' = \text{proj}_{\pi'}c$; however $\rho(c) = c$, and for $z \in St(\pi)$ if $\rho(z) = \text{proj}_{\pi'}c$, then $z = \text{proj}_{\pi}c$. Therefore $x = \text{proj}_{\pi}c = y$, a contradiction.

Step 4. If $x' = y'$, then $\ell(w(y)) > \ell(w(x))$.

Since x is nearer the vertex of T than y is, the sector direction $T(x)$ points towards π rather than away from it. Thus $\rho(T(x))$ is on the $-\alpha$ side of M, and by Step 3 this is true of $S(\rho(x))$ also (see Figure 9.4). In other words the elements $w(x)$ and 1 of W_o lie on one side of a wall, and by Step 2 $w(y) = rw(x)$ lies on the opposite side of the wall, hence $\ell(w(y)) > \ell(w(x))$.

We now define T_1 to be any subsector of T with base chamber x_o such that $w(x_o)$ has maximal length in W_o. Using Step 4 we see that if x and y are adjacent and distinct chambers of T_1, then $\rho(x) \neq \rho(y)$, completing the proof. $\square$

Proof of (9.5). We have two sectors S and T, and wish to find subsectors $S_1 \subset S$ and $T_1 \subset T$ lying in a common apartment.

Let A be an apartment containing S, set $\rho = \rho_{S,A}$ and let $T_1 \subset T$ be a subsector as in (9.9), having base chamber x_o, and such that $\rho|T_1$ is an isometry. If S' denotes the translate of S in A having the same vertex as $\rho(T_1)$, then we let $S_1 \subset S \cap S'$ be a subsector lying in a common apartment with x_o. It suffices to show that for all chambers $c \in S_1$, $\rho_{c,A}|T_1 = \rho|T_1$.

Given $c \in S_1$ and $y \in T_1$ we work by induction along a minimal gallery from x_o to y. Since S_1 and x_o lie in a common apartment, the induction can start. Now as in the proof of (9.9) let $x, y \in T_1$ be distinct chambers on a common panel π, and with x closer to x_o; again write $x' = \rho(x)$, $y' = \rho(y)$, $\pi' = \rho(\pi)$, and let α be the root of A containing x' but not y'. By induction $\rho_{c,A}(x) = x'$, and we must show $\rho_{c,A}(y) = y'$.

If $c \in \alpha$, then $x' = \text{proj}_{\pi'}c$, and so $x = \text{proj}_{\pi}c$ is the unique chamber of $St(\pi)$ mapped onto x' by $\rho_{c,A}$; therefore $\rho_{c,A}(y) \neq x'$, and hence $\rho_{c,A}(y) = y'$ as required.

If $c \in -\alpha$, then $y' = \text{proj}_{\pi'}c$, and it suffices to show that $y = \text{proj}_{\pi}c$. We first claim that $\rho|St(\pi) = \rho_{e,A}|St(\pi)$ for some chamber $e \in -\alpha$. Indeed S_1 contains a subsector lying entirely in $-\alpha$ (because it lies in a sector having vertex $\rho(T_1) \in \alpha$, and has a chamber $c \in -\alpha$), hence contains a chamber e as required. Thus $\rho_{e,A}(\text{proj}_{\pi}c) = \rho(\text{proj}_{\pi}c) = \text{proj}_{\pi'}c = y' = \rho_{e,A}(y)$. Since $y' = \text{proj}_{\pi'}e$ its inverse image under $\rho_{e,A}$ is $\text{proj}_{\pi}e$, so we have $\text{proj}_{\pi}c = \text{proj}_{\pi}e = y$, as required. $\square$

Notes. Tits systems (B, N) of affine type were introduced by Iwahori and Matsumoto [1965], and in the literature the terms "Iwahori subgroup" for B, and "parahoric subgroups" for the P_J, are often used. The general theory of affine buildings, and the construction of the building at infinity, is developed by Bruhat and Tits [1972]; their work includes a description of the building as a metric space - see also Brown [1979] Ch. VI, and "non-discrete buildings" obtained from non-discrete valuations - see Appendix 3. The proof of (9.5) is taken from [loc. cit.] 2.9.5 (pages 58-60).

Exercises to Chapter 9

1. In a given affine Coxeter complex, let $\alpha_1, \ldots, \alpha_t$ be roots whose walls are linearly independent, and for each $i = 1, \ldots, t$ let β_i be a translate of α_i. Prove that there is a translation g (not necessarily in W) such that $\beta_i = g(\alpha_i)$ for all $i = 1, \ldots, t$.

2. If S' is a translate of a sector S in an affine Coxeter complex, show that $S \cap S'$ is a sector. [HINT: Use Exercise 1].

3. Show that two sectors are parallel if and only if their intersection contains a sector.

4. Given a chamber c, and a half-apartment (root) α, is there necessarily a half-apartment $\beta \subset \alpha$ such that c and β lie in a common apartment? (cf. 9.4).

5. Describe the triangulations of $\mathbf{R}^3$ determined by the Coxeter complexes of types $\widetilde{C}_3$, $\widetilde{B}_3$ and $\widetilde{A}_3$.

6. Show that the map, in the early part of section 1, from the Coxeter complex W onto $\mathbf{R}^n$ is a local homeomorphism. [Hint: For each simplex σ consider this map restricted to $St(\sigma)$, and work by induction on the codimension of σ].

7. Show that if a sector-face contains two sector-faces of the same dimension, then their intersection is also a sector-face of that dimension. (cf. 9.2 for sectors).

8. Show that in an affine Coxeter complex a convex set of chambers is closed and convex in the Euclidean sense. [HINT: A convex set of chambers is either the whole Coxeter complex, or is an intersection of roots, by (2.8)].

9. Show that the intersection of two apartments is closed and convex in both. [HINT: If p and q are points of A and A', let $x \in A$ and $y \in A'$ be chambers containing p and q respectively, and let A'' be an apartment containing x and y: cf. Exercise 6 in Chapter 3].

Exercises 10-12 deal with the $\widetilde{A}_{n-1}$ example of section 2.

10. If L and L' are $\mathcal{O}$-lattices show there is a basis $e_1, \ldots, e_n$ for L such that L' is spanned by $\pi^{r_1} e_1, \ldots, \pi^{r_n} e_n$. Conclude that any two vertices lie in an apartment determined by a basis of V.

11. Define a circuit of vertices and edges to be *minimal* if it contains a path of shortest length joining any two of its vertices. Show that a minimal circuit lies in an apartment determined by a basis of V. Use this to show Δ is simply-connected in the topological sense.

12. Show that if a group of automorphisms fixes two vertices x and y, then it fixes something in $St(x)$. [HINT: Consider an apartment containing x and y].

Chapter 10
AFFINE BUILDINGS II

This chapter deals with the relationship between an affine building Δ having a system of apartments $\mathcal{A}$, and its spherical building at infinity denoted $(\Delta, \mathcal{A})^\infty$. When this building at infinity is Moufang (e.g. whenever Δ has rank at least 4), one obtains root groups with a valuation (section 3), which are then used in section 4 to recover $(\Delta, \mathcal{A})$, and assist in the classification (section 5). An application to finite group theory is given in section 6.

As a matter of notation the term *root* will be reserved for spherical buildings such as $(\Delta, \mathcal{A})^\infty$, and we use Latin letters $a, b, c, \ldots$ for such roots. A root of an affine building will be called a *half-apartment* or *affine root*, and we use Greek letters $\alpha, \beta, \gamma, \ldots$ for these.

1. Apartment Systems, Trees and Projective Valuations.

Given an affine building Δ, an *apartment system* for Δ (or more precisely a *discrete* apartment system - cf. Appendix 3) will mean that a set $\mathcal{A}$ of apartments of Δ is given, satisfying (i) and (ii) below. This data will be referred to as $(\Delta, \mathcal{A})$, and a *sector-face*, or a *wall*, of $(\Delta, \mathcal{A})$ will mean a sector-face, or a wall, in some apartment of $\mathcal{A}$. The conditions are:

(i) every chamber lies in some apartment of $\mathcal{A}$.

(ii) any two sectors of $(\Delta, \mathcal{A})$ contain subsectors lying in a common apartment of $\mathcal{A}$.

For example if $\mathcal{A}$ is the set of all apartments of Δ, then by (3.6) and (9.5) both (i) and (ii) hold, and in this case $\mathcal{A}$ is called *complete*.

The Building at Infinity. For any apartment system $\mathcal{A}$, the parallel classes of sector-faces are the simplexes of a *building at infinity*, which we denote $(\Delta, \mathcal{A})^\infty$. To see this notice that $(\Delta, \mathcal{A})^\infty$ is a subcomplex of the

Δ^∞ of Chapter 9, and by condition (ii) any two chambers of $(\Delta, \mathcal{A})^\infty$ lie in a common apartment A^∞ where $A \in \mathcal{A}$. Therefore by (3.11), or Exercise 10 of Chapter 3, $(\Delta, \mathcal{A})^\infty$ is a sub-building of Δ^∞, and its apartments are the A^∞ for $A \in \mathcal{A}$. If X is a sector-face or wall of $(\Delta, \mathcal{A})$, then as before we let $x = X^\infty$ be the simplex or wall it determines "at infinity" in $(\Delta, \mathcal{A})^\infty$, and we say X has *direction* x.

Example 1. As in Chapter 9 section 2, let K be a field with a discrete valuation v, and $\widehat{K}$ its completion with respect to v. Let V be an n-dimensional vector space over K, let $\widehat{V} = V \otimes_K \widehat{K}$, and let $\mathcal{O}$ and $\widehat{\mathcal{O}}$ be the valuation rings of K and $\widehat{K}$. The building $\widetilde{A}_{n-1}(K, v)$ (or $\widetilde{A}_{n-1}(\widehat{K}, v)$) has as its vertices the equivalence classes [L] of $\mathcal{O}$-lattices (or $\widehat{\mathcal{O}}$-lattices) in V (or $\widehat{V}$), under the equivalence relation $[L] = [L'] \Leftrightarrow L = aL'$ for some $a \in K$ (or $\widehat{K}$); these buildings are isomorphic as chamber systems. Let Δ denote this common building; it acquires a system of apartments $\mathcal{A}(K)$ or $\mathcal{A}(\widehat{K})$ (as in Chapter 9 section 2), by taking decompositions of V or $\widehat{V}$ respectively into 1-spaces $\langle e_1 \rangle \oplus \ldots \oplus \langle e_n \rangle$. In fact $\mathcal{A}(\widehat{K})$ is the complete system of apartments - see Exercises 1 and 2. Thus the building at infinity Δ^∞ of Chapter 9, obtained by using all possible apartments, is the $A_{n-1}(\widehat{K})$ building, whereas $(\Delta, \mathcal{A}(K))^\infty$ is the $A_{n-1}(K)$ building.

Trees with sap - the rank 2 case. An affine building of rank 2 is a tree with no end points (Exercise 12 of Chapter 3), and if an apartment system is specified, we shall call it a *tree with sap*. Its *ends* are the parallel classes of sectors; there are no sector-panels, and $(\Delta, \mathcal{A})^\infty$ is just the set of ends - a rank 1 building of spherical type.

Example 2. $SL_2(K)$. Let $n = 2$ in Example 1; the building $\widetilde{A}_1(K, v)$ is a tree with sap, whose ends are the 1-spaces $\langle v \rangle \subset V$. Sectors (i.e. half-apartments) having $\langle v \rangle$ as an end are given by sequences of lattices:

$$L_n = \pi^n L_o + (L_o \cap \langle v \rangle)$$

where π is a uniformizer (i.e., π generates the maximal ideal of $\mathcal{O}$). $\square$

By definition two distinct ends a and b of a tree with sap lie in a common apartment of $\mathcal{A}$, which is obviously unique; we denote it $[a, b]$. Moreover three distinct ends a, b, c determine a unique *junction* $\kappa(a, b, c)$ (French: *carrefour*), the vertex common to $[a, b]$, $[b, c]$ and $[c, a]$.

We assume our trees with sap are endowed with a metric, in other words a distance between any two vertices x and y, equal to the sum of the distances between adjacent vertices on the path from x to y. The trees in section 2, which are obtained from affine buildings of higher rank, will come equipped with a metric induced from a metric on Euclidean space, and for this case the distance between any pair of adjacent vertices is a constant.

Now given four distinct ends a, b, c, d we let $\omega(a, b; c, d)$ denote the distance from $\kappa(a, b, c)$ to $\kappa(a, b, d)$, in the direction from a to b (i.e., with a $+$ or $-$ sign according to whether $\kappa(a, b, c)$ precedes or follows $\kappa(a, b, d)$ in the line from a to b) - see Figure 10.1.

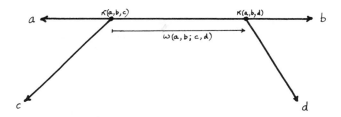

Figure 10.1

(10.1) LEMMA. *The function ω satisfies:*
(ω1) $\omega(a, b; c, d) = \omega(c, d; b, a) = -\omega(a, b; d, c)$,
(ω2) *if* $\omega(a, b; c, d) = k > 0$, *then* $\omega(a, d; c, b) = k$ *and* $\omega(a, c; b, d) = 0$,
(ω3) $\omega(a, b; c, d) + \omega(a, b; d, e) = \omega(a, b; c, e)$.

PROOF: Exercise.   □

Any function ω taking values in $\mathbf{R}$ and satisfying (ω1), (ω2) and (ω3) is called a *projective valuation*. If it takes values in a discrete subset of $\mathbf{R}$ we call it *discrete*.

(10.2) THEOREM. *Let $(T, \mathcal{A})$ be a tree with sap in which each vertex lies on at least three edges, and let ω be the projective valuation on $(T, \mathcal{A})^\infty$. Then ω determines $(T, \mathcal{A})$ up to unique isomorphism.*

PROOF: Exercise.   □

Notice that if we did not require each vertex to have valency at least three, we could subdivide T, for instance by inserting a vertex in the middle of each edge, to obtain the same $(T, \mathcal{A})^\infty$ and ω. However, using a more general notion of "tree" as a union of copies of $\mathbf{R}$, vertices no longer exist as such, and (10.2) can be greatly strengthened (see (A.16) in Appendix

3) to say that if ω is a projective valuation on a set having more than two elements, then it arises from such a "tree" which is determined up to unique isomorphism.

Example 3. Let K be a field, and ω a projective valuation on the set $K \cup \{\infty\}$, invariant under the affine group $\{x \mapsto ax + b \mid a \in K^{\times}, b \in K\}$. Define $v : K \to \mathbf{R} \cup \{\infty\}$ by:

$$v(x) = \omega(\infty, 0; 1, x) \text{ for } x \neq 0, 1$$
$$v(0) = \infty$$
$$v(1) = 0$$

We will show that v is a (rank 1) valuation in the usual sense, namely that
$$v(ab) = v(a) + v(b)$$
and
$$v(a + b) \geq \min(v(a), v(b)).$$

We first observe that invariance under the affine group implies:

$$\omega(\infty, b; c, d) = \omega(\infty, 0; (c - b), (d - b)) = v((c - b)^{-1}(d - b)).$$

Thus

$$\begin{aligned}
v(ab) &= \omega(\infty, 0; 1, ab) \\
&= \omega(\infty, 0; a^{-1}, b) \\
&= \omega(\infty, 0; a^{-1}, 1) + \omega(\infty, 0; 1, b) \\
&= v(a) + v(b)
\end{aligned}$$

Now suppose, by way of contradiction, that $v(a + b) < v(a), v(b)$. Since

$$\begin{aligned}
v(a + b) &= \omega(\infty, 0; 1, a + b) \\
&= \omega(\infty, 0; 1, a) + \omega(\infty, 0; a, a + b) \\
&= v(a) + \omega(\infty, -a; 0, b),
\end{aligned}$$

we have $\omega(\infty, -a; 0, b) < 0$, hence $\omega(\infty, -a; b, 0) > 0$. By $(\omega 3)$ this implies $\omega(\infty, 0; b, -a) > 0$, and hence $v(-b^{-1}a) > 0$. Similarly, interchanging a and b, we have $v(-a^{-1}b) > 0$. Thus $v(1) > 0$, a contradiction; hence $v(a + b) \geq \min(v(a), v(b))$ as required.

Finally we remark that by using ($\omega 2$) it is straightforward to show that:

$$\omega(a,b;c,d) = v((d-1)^{-1}(c-a)(c-b)^{-1}(d-b)),$$

and it follows from this that the invariance of ω under the affine group implies its invariance under the projective group - see Exercise 3.

2. Trees associated to Walls and Panels at Infinity.

Given a wall m of the building at infinity $(\Delta, \mathcal{A})^\infty$, we shall define a tree with sap $T(m)$ whose ends are the roots of $(\Delta, \mathcal{A})^\infty$ containing m; this set of roots will be denoted $St(m)$. Similarly, given a panel π of $(\Delta, \mathcal{A})^\infty$ we shall define $T(\pi)$, a tree with sap, whose ends are the chambers containing π, this set of chambers being denoted as usual by $St(\pi)$. If π is contained in m, there is a canonical isomorphism from $T(\pi)$ to $T(m)$. This induces a bijection from $St(\pi)$ to $St(m)$ which associates to each chamber x of $St(\pi)$ the unique root having wall m and containing x - see (6.3).

The tree $T(m)$ with sap. For a given wall m of $(\Delta, \mathcal{A})^\infty$, the vertices of $T(m)$ are walls M of $(\Delta, \mathcal{A})$ such that $M^\infty = m$, and two vertices are joined by an edge if they are walls of a common apartment with no wall in between. The apartments of $T(m)$ are taken to consist of those vertices, and edges joining them, which are walls of some common apartment in $\mathcal{A}$. The half-apartments (i.e. sectors) of $T(m)$ then correspond in an obvious way to those half-apartments of $(\Delta, \mathcal{A})$ whose boundary wall has direction m. Thus the ends of $T(m)$ are simply the roots of $(\Delta, \mathcal{A})^\infty$ having boundary wall m.

Given two distinct roots of a spherical building $((\Delta, \mathcal{A})^\infty$ in our case) having a common boundary wall, there is a unique apartment containing them both - cf. (6.3). Thus each pair of ends of $T(m)$ determines an apartment of $(\Delta, \mathcal{A})^\infty$, hence an apartment of $(\Delta, \mathcal{A})$, and hence an apartment of $T(m)$ itself. We have therefore proved:

(10.3) LEMMA. $T(m)$ *is a tree with sap whose ends correspond to the roots of* $(\Delta, \mathcal{A})^\infty$ *having boundary wall* m *(i.e., to* $St(m)$*).*   □

Before dealing with $T(\pi)$, we define two sector-panels to be *asymptotic* if their intersection contains a sector-panel. By Exercise 5 this is an equivalence relation which is finer than the relation of being parallel. The equivalence classes will be called *asymptote classes*, and the asymptote class of D will be denoted $\widehat{D}$.

The tree $T(\pi)$ with sap. For a given panel π of $(\Delta, \mathcal{A})^\infty$ the vertices of $T(\pi)$ are the asymptote classes of sector-panels D for which $D^\infty = \pi$, and two vertices are joined by an edge if there are sector-panels from the two classes, lying in the same sector, and with no sector-panel in between. Before defining the apartments of $T(\pi)$ we observe that if $D^\infty = \pi$, and S is any sector of $(\Delta, \mathcal{A})$ having D as a sector-panel, then the other sector panels parallel to D and contained in S form a half-line in $T(\pi)$ - see Figure 10.2.

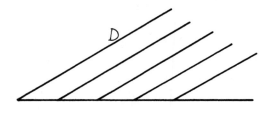

Figure 10.2

We define the apartments of $T(\pi)$ by requiring these half-lines to be the half-apartments (i.e., sectors) of $T(\pi)$. Thus if two sectors S and T lie in a common apartment and intersect in the sector-panel D, then the sector-panels of $S \cup T$ parallel to D form an apartment of $T(\pi)$. Since any two sectors of $(\Delta, \mathcal{A})$ contain subsectors lying in a common apartment, the same is true for half-apartments of $T(\pi)$, and we have a tree with sap. Moreover two half-apartments of $T(\pi)$ have the same end if and only if the corresponding sectors S and S' contain a common sector (i.e., S and S' give the same chamber $S^\infty = (S')^\infty$ of $(\Delta, \mathcal{A})^\infty$). Thus the ends of $T(\pi)$ correspond to the chambers of $St(\pi)$, and we have proved:

(10.4) LEMMA. $T(\pi)$ is a tree with sap whose ends correspond to the chambers of $(\Delta, \mathcal{A})^\infty$ having π as a panel (i.e., chambers of $St(\pi)$). □

The idea is now to use section 1, applied to $T(m)$ and $T(\pi)$, to obtain projective valuations ω_m and ω_π on $St(m)$ and $St(\pi)$ respectively. This requires a metric on $T(m)$ and $T(\pi)$ which we define as follows. The affine Coxeter complex, regarded as Euclidean space, can be given a metric (unique up to multiplication by a positive real number). This gives a metric on each apartment, and hence on $(\Delta, \mathcal{A})$, so we have a distance between any two parallel walls or sector-panels, which in turn defines a metric on $T(m)$ and $T(\pi)$.

Remark. The distance between adjacent vertices of $T(m)$ or $T(\pi)$ cannot be 1 in all cases. In fact if the Coxeter group has two orbits on the set of walls, then the ratio of the distances between adjacent walls in the two orbits is $\sqrt{2}$ (in the $\widetilde{B}_n$, $\widetilde{C}_n$ and $\widetilde{F}_4$ cases) or $\sqrt{3}$ (in the $\widetilde{G}_2$ case). For example Figure 10.3 shows the $\widetilde{C}_2$ case.

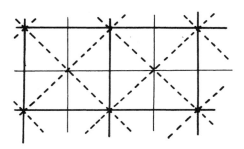

Figure 10.3

We now choose a fixed Euclidean metric, and let ω_m and ω_π be the projective valuations induced on $St(m)$ and $St(\pi)$.

The following theorem is proved by Tits [1986a] section 18, but in a more general setting in which the "building" may not be discrete - see Appendix 3. We shall not prove it here.

(10.5) THE UNIQUENESS THEOREM. *If $(\Delta, \mathcal{A})^\infty$ is thick, then $(\Delta, \mathcal{A})$ is determined up to unique isomorphism by the ω_π (or the ω_m) for all panels π (or walls m) of $(\Delta, \mathcal{A})^\infty$.* $\square$

The data ω_π for all panels π can in fact be inferred from knowing just one or two of the ω_π, namely one in each of the one or two types of walls of $(\Delta, \mathcal{A})^\infty$. The idea is that one can transfer the data ω_π to the data $\omega_{\pi'}$ whenever π and π' lie in a common wall. The following proposition makes this precise.

(10.6) PROPOSITION. *If π is a panel in a wall m, then for each asymptote class $\widehat{D}$ of sector-panels in the direction π, there is a unique wall M in the direction m containing a representative of $\widehat{D}$. The map $\widehat{D} \mapsto M$ is an isomorphism from $T(\pi)$ to $T(m)$ and induces on the set of ends a map $\iota_{\pi m}$ sending a chamber of $St(\pi)$ to the unique root of $St(m)$ containing it. Moreover $\omega_\pi = \omega_m \circ \iota_{\pi m}$.*

PROOF: Let D_1 be any sector-panel in the direction π. Take an apartment $A_1 \in \mathcal{A}$ which contains D_1, and let S_1 and T_1 be the distinct sectors of A_1 having D_1 as a face. The chambers S_1^∞ and T_1^∞ each have a panel, namely π, in m, and therefore by (6.3) S_1^∞, T_1^∞ and m lie in a unique common apartment A^∞.

In A there are subsectors S and T of S_1 and T_1 respectively, and since the convex hull of S and T is the same in both A_1 and A, it contains a subsector D of D_1 - see Figure 10.4.

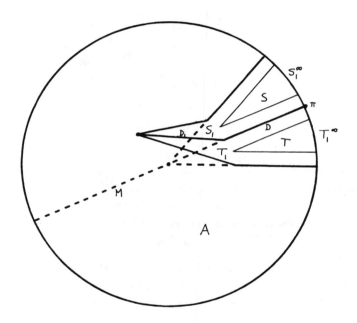

Figure 10.4

Let M be the unique wall of A containing D; then M^∞ is the wall of A^∞ containing $D^\infty = \pi$, and hence $M^\infty = m$.

Given $\hat{D}$, the uniqueness of M is an immediate consequence of the fact that two parallel walls are either equal or disjoint. Moreover the map $\hat{D} \mapsto M$ is a bijection since all sector-panels of M in the direction π are obviously asymptotic.

The remainder of the proof is straightforward and is left as an exercise.

□

If π and π' are two panels of m, then $\iota_{\pi'm}^{-1} \circ \iota_{\pi m}$ is a bijection from $St(\pi)$ to $St(\pi')$; any combination of such bijections is called a *projectivity*, and we let $GP(\pi)$ denote the group of projectivities from $St(\pi)$ to itself. By (10.6) any projectivity from π to π' sends ω_π to $\omega_{\pi'}$; in particular ω_π is invariant under the group $GP(\pi)$.

In a spherical Coxeter complex (with a connected diagram) there are at most two types of walls (this was discussed earlier in Chapter 6 section 4). Moreover in a thick building of spherical type, given two panels π_1 and π_2 of the same type, there is a third panel π' opposite both of them (it is an exercise to verify this - cf. (3.3) Step 2). Hence π_1 lies in a common wall with π' which in turn lies in a common wall with π_2. Therefore in $(\Delta, \mathcal{A})^\infty$ there are at most two "projectivity classes" of panels, and we have the following.

(10.7) COROLLARY TO THE UNIQUENESS THEOREM. *If Δ is thick, then $(\Delta, \mathcal{A})$ is determined up to unique isomorphism by ω_π for a single panel π in one of at most two "projectivity classes".*

Application - the classification of $\widetilde{A}_n$ buildings for $n \geq 3$. Let Δ be an affine building of type $\widetilde{A}_n$ for $n \geq 3$, and let $\mathcal{A}$ be a system of apartments for Δ. Then $(\Delta, \mathcal{A})^\infty$ is a spherical building of type A_n, and since $n \geq 3$ it is the $A_n(K)$ building for some field K (not necessarily commutative). If π is a panel of $(\Delta, \mathcal{A})^\infty$, then $St(\pi)$ can be identified with the projective line $K \cup \{\infty\}$, and $GP(\pi) \cong PGL_2(K)$. It therefore follows from Example 3 in section 1 that ω_π is equivalent to a discrete valuation v of K. The same building at infinity with the same ω_π could be obtained by using the $\widetilde{A}_n(K, v)$ building of Chapter 9 section 2, and so (10.7) implies that $(\Delta, \mathcal{A}) \cong \widetilde{A}_n(K, v)$. (The isomorphism is uniquely determined by the isomorphism of the buildings at infinity).

This argument applies to other types of affine buildings, namely those of types $\widetilde{D}_n$, $\widetilde{E}_6$, $\widetilde{E}_7$, $\widetilde{E}_8$, where there is only one "projectivity class" of panels, and $St(\pi)$ is a projective line; it shows that every such building is associated to a field K with a discrete valuation v. However we have not yet constructed these affine buildings, and do not therefore yet have a classification. In the next section we study "root groups with a valuation" which we use in section 4 to construct affine buildings. In section 5 we then use K and v to obtain a system of "root groups with valuation" to conclude the classification of these cases.

3. Root Groups with a valuation.

In this section we assume $(\Delta, \mathcal{A})^\infty$ to be Moufang. By (6.7) this is always the case when $(\Delta, \mathcal{A})^\infty$ has rank at least 3 (recall we assume a connected diagram for $(\Delta, \mathcal{A})$, and hence also for $(\Delta, \mathcal{A})^\infty$). Now fix some apartment $A \in \mathcal{A}$, and let Φ be the set of roots of A^∞. For each $a \in \Phi$ let U_a denote the corresponding root group.

(10.8) PROPOSITION. *If G is the group generated by the U_a, the action of G on $(\Delta, \mathcal{A})^\infty$ extends to an action on $(\Delta, \mathcal{A})$.*

PROOF: If U_a is any root group we must show that its action extends to $(\Delta, \mathcal{A})$. By (10.7) it suffices to check that U_a preserves ω_π for π in one or two possible classes. In either case π may be taken to lie in $a - \partial a$, so U_a acts trivially on $St(\pi)$, and hence on ω_π.   □

We now fix a point $s \in A$ (not necessarily a vertex). Each root $a \in \Phi$ corresponds to a half-space a_s of A, having s on its boundary (if we treat A as a vector space V_s with origin s, each a_s is a half-space of V_s as in Chapter 2 section 4). Now given $u \in U_a - \{1\}$, $A \cap uA$ is a half-apartment of A, and its boundary wall M_u is parallel to ∂a_s. We define $\varphi_a(u)$ to be the distance from ∂a_s to M_u in the Euclidean space A, measured in the $+a_s$ to $-a_s$ direction (i.e. with a $+$ sign if $\partial a_s \subset uA$, and a $-$ sign otherwise) - see Figure 10.5 where $\varphi_a(u)$ is negative.

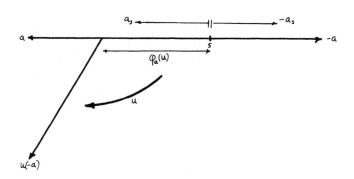

Figure 10.5

As mentioned in section 2 we do not have a metric defined on A *a priori*; it is only unique up to a multiplicative constant. After fixing this metric on A, each root $a \in \Phi$ determines φ_a up to an additive constant depending on the choice of the point s.

Since $(\Delta, A)^\infty$ is Moufang, U_a acts simple-transitively on the set of apartments of $(\Delta, A)^\infty$ containing a, and so for $m = \partial a$, U_a corresponds to the set of roots of $St(m)$ different from a (see (6.3)). Thus $St(m)$ consists of a and $u(-a)$ as u ranges over U_a.

(10.9) LEMMA. $\omega_m(a, u(-a); u'(-a), u''(-a)) = \varphi_a(u^{-1}u'') - \varphi_a(u^{-1}u')$.

PROOF: By (10.8) u^{-1} fixes ω_m, so the left hand side equals $\omega_m(a, -a; u^{-1}u'(-a), u^{-1}u''(-a))$. The result follows immediately from the definitions of φ_a and ω_m. □

Root Data with Valuation. (Données radicielles valuées).

As in Chapter 6 let Φ be the set of roots in an apartment Σ of a (thick) Moufang building of spherical type, with a connected diagram, and for each $a \in \Phi$ let U_a denote the corresponding root group. As mentioned in Chapter 2 section 4 we may regard roots as half-spaces, and walls ∂a as hyperplanes, in a real vector space V (the Coxeter group W acts on V preserving a dot product). We let 1_a denote the vector of length 1 perpendicular to ∂a and contained in a, and let r_a denote the reflection in the wall ∂a, switching a and $-a$.

A collection $\psi = (\psi_a)_{a \in \Phi}$ of maps $\psi_a : U_a \to \mathbf{R}$ will be called a *valuation* of the U_a if they satisfy the following.

(V0) Card $\psi_a(U_a) \geq 3$

(V1) $U_{a,t} := \psi_a^{-1}[t, \infty]$ is a group, and $U_{a,\infty} = \{1\}$.

(V2) Given $b \neq \pm a$, the commutator

$$[U_{a,k}, U_{b,\ell}] \leq \langle U_{c,pk+q\ell} \mid c \in (a, b)\rangle$$

where $1_c = p \, 1_a + q \, 1_b$ (recall from Chapter 6 section 3 that (a, b) is the set of roots c with $a \cap b \subset c \neq a, b$).

(V3) Given $a, b \in \Phi$, and $u \in U_a - \{1\}$, there exists $t \in \mathbf{R}$ (depending on b and u) such that for all $x \in U_b$

$$\psi_{r_a(b)}(m(u)xm(u)^{-1}) = \psi_b(x) + t.$$

Moreover if $a = b$, then $t = -2\psi_a(u)$, or in other words

$$\psi_{-a}(m(u)xm(u)^{-1}) = \psi_a(x) - 2\psi_a(u).$$

The element $m(u)$ was defined in Chapter 6 section 4; it stabilizes Σ, switching a with $-a$. It is the unique element $vuv' \in U_{-a}uU_{-a} \cap N$.

(10.10) EXERCISE. *Given* $m(u) = vuv'$ *then* $m(v) = m(u) = m(v')$ *by (A.1) in Appendix 1. Use this to prove:*

(a) $\psi_{-a}(v) = \psi_{-a}(v')$

(b) $\psi_a(u) = -\psi_{-a}(v)$.

[HINT: (a) is immediate from $m(v) = m(v')$ *and (V3); for (b) let* $m(v) = u_1 v u$, $m(u_1) = v_1 u_1 v$, *so* $m(u_1)um(u_1)^{-1} = v_1$ *- now use (a).]*

Remark. We have not assumed that $\psi_a(U_a - \{1\})$ is a discrete subset of **R**; if it is we call the valuation *discrete*. These are the valuations that arise from affine buildings, and in section 4 we shall use such valuations to construct an affine BN-Pair. However, even in the non-discrete case it is possible to construct a geometry having "at infinity" the Moufang building for the root groups U_a; these "non-discrete buildings" are discussed in Appendix 3.

Example 4. Let $\Phi = \{\pm a, \pm b, \pm c\}$ be the roots in an A_2 apartment in such a way that $1_c = 1_a + 1_b$, as shown.

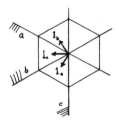

Let K be a field with a valuation v satisfying the conditions of Chapter 9 section 2, though v need not be discrete. Then v determines a valuation of the root groups in $SL_3(K)$ as follows. After choosing a suitable basis we may write

$$U_{a,k} : \begin{pmatrix} 1 & x & 0 \\ 0 & 1 & 0 \\ 0 & 0 & 1 \end{pmatrix} \quad \text{where } v(x) \geq k$$

$$U_{b,\ell} : \begin{pmatrix} 1 & 0 & 0 \\ 0 & 1 & y \\ 0 & 0 & 1 \end{pmatrix} \quad \text{where } v(y) \geq \ell$$

$$U_{c,m} : \begin{pmatrix} 1 & 0 & z \\ 0 & 1 & 0 \\ 0 & 0 & 1 \end{pmatrix} \quad \text{where } u(z) \geq m$$

If $g = \begin{pmatrix} 1 & x & 0 \\ 0 & 1 & 0 \\ 0 & 0 & 1 \end{pmatrix}$ and $h = \begin{pmatrix} 1 & 0 & 0 \\ 0 & 1 & y \\ 0 & 0 & 1 \end{pmatrix}$ then $[g,h] = \begin{pmatrix} 1 & 0 & xy \\ 0 & 1 & 0 \\ 0 & 0 & 1 \end{pmatrix}$. Since $v(xy) = v(x) + v(y)$, this gives $[U_{a,k}, U_{b,\ell}] = U_{c,m}$ where $m = k + \ell$ (a case where $p = q = 1$). $\qquad\qquad\qquad\qquad\qquad\qquad\qquad\qquad\qquad\qquad\qquad\qquad$ □

(10.11) THEOREM. *The (φ_a) defined earlier, using a point s in the affine apartment A, are a valuation of the root groups (U_a).*

PROOF: Since $(\Delta, \mathcal{A})$ is thick, φ_a takes infinitely many values, so (V0) is clear.

(V1): Notice first that $uA \cap A = u^{-1}A \cap A$, hence $\varphi_a(u) = \varphi_a(u^{-1})$, and so $U_{a,t}$ contains the inverse of each of its elements. Furthermore for $u, u' \in U_a$ the part of A fixed by u and u' is certainly fixed by uu', so $uA \cap u'A \cap A \subset uu'A \cap A$; hence $\varphi_a(uu') \geq \min(\varphi_a(u), \varphi_a(u'))$, and $U_{a,t}$ is a group. Moreover if $u \in U_{a,\infty}$, then u fixes A and hence $u = 1$.

(V2): Let K and L be the walls parallel to, and at distance k and ℓ from, ∂a_s and ∂b_s respectively - see Figure 10.6.

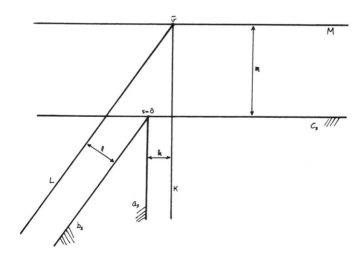

Figure 10.6

We now consider the underlying affine space A as a vector space with origin s, and let $\bar{v}$ denote the point of intersection of L and K in the 2-space spanned by 1_a and 1_b. By (6.12) $[U_a, U_b] \leq \langle U_c \mid c \in (a,b) \rangle$. If $u \in U_c$

is a factor in such a commutator then u must fix $\bar{v}$, and hence $u \in U_{c,m}$ where m is the (signed) distance from ∂c_s to $\bar{v}$ (see Figure 10.6). We let M denote the wall through $\bar{v}$ parallel to ∂c_s so m is the distance from ∂c_s to M. To evaluate m, notice that the equations for points $\bar{x}$ on the walls K, L and M are respectively:

$$\bar{x} \cdot 1_a = -k$$
$$\bar{x} \cdot 1_b = -\ell$$
$$\bar{x} \cdot 1_c = -m$$

As $\bar{v}$ lies on K, L and M this gives

$$m = -\bar{v} \cdot 1_c = -p\bar{v} \cdot 1_a - q\bar{v} \cdot 1_b = pk + q\ell$$

where $1_c = p1_a + q1_b$.

(V3) Given $a, b \in \Phi$ write $b' = r_a(b)$. For $u \in U_a - \{1\}$, M_u is the wall fixed by $m(u)$, and we let t be the distance from $M_u \cap \partial b_s$ to $\partial b'_s$ (in the b'_s to $-b'_s$ direction) - see Figure 10.7.

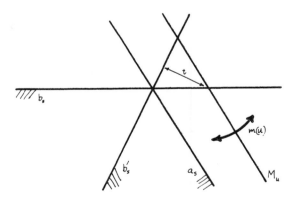

Figure 10.7

Then using signed distances as usual, $m(u)$ sends a wall at distance k from ∂b_s to a wall at distance $k+t$ from $\partial b'_s$. Thus $\varphi_{b'}(m(u)xm(u)^{-1}) = \varphi_b(x)+t$ for all $x \in U_b$, as required.

Finally if $a = b$, then $b_s = -b'_s$, and we let h be the distance from $\partial b'_s$ to M_u (+ if $M_u \subset b'_s$, and − if $M_u \subset b_s$) - see Figure 10.8.

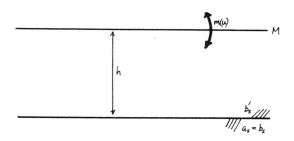

Figure 10.8

Then $m(u)$ sends a wall at distance k from ∂a_s to a wall at distance $k + 2h$ from ∂a_s, so $t = 2h = -2\varphi_a(u)$ in this case. □

Equivalence and equipollence. To conclude this section, replace the point s by s', and keep the same metric on A. If v is the vector $s' - s$, and φ' the valuation obtained using s', then

$$\varphi'_a(u) = \varphi_a(u) + v \cdot 1_a$$

- see Figure 10.9.

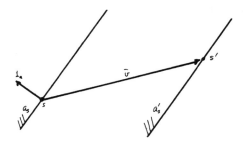

Figure 10.9

Two valuations related in this way are called *equipollent*, and we write

$$\varphi' = \varphi + v.$$

Altering the metric on A turns $\varphi_a(u)$ into $\lambda\varphi_a(u)$ for some positive real number λ, and if we alter both the metric and the point s then in place of φ we obtain $\lambda\varphi + v$, for some $\lambda > 0$ and $v \in V$, where

$$(\lambda\varphi + v)_a(u) = \lambda\varphi_a(u) + v \cdot 1_a.$$

We call φ and $\lambda\varphi + v$ *equivalent*. If ψ is any valuation (satisfying (V0) - (V3)), then so is $\lambda\psi + v$ (Exercise 7).

4. Construction of an Affine BN-Pair.

In this section we start with a discrete valuation $\psi = (\psi_a)$ of the root groups $(U_a)_{a \in \Phi}$ as defined in section 3, and construct an affine BN-Pair. We let N be the subgroup generated by the $m(u)$, and show that it acts as an affine Coxeter group W^{aff} on the affine space whose points are valuations equipollent to one another. The finite Coxeter group acting on Φ will be denoted $W(\Phi)$.

Recall again from Chapter 2 section 4 that $W(\Phi)$ acts on a real vector space V preserving a dot product. The roots $a \in \Phi$ correspond to half-spaces of V and the walls ∂a to hyperplanes of V. As in section 3 we let 1_a denote the unit vector perpendicular to ∂a and lying in the half-space a.

For $n \in N$ with image $w \in W(\Phi)$ we define an action on the set of valuations, as follows:

$$(n \cdot \psi)_a(u) = \psi_{w^{-1}(a)}(n^{-1}un).$$

In fact $n \cdot \psi$ is a valuation equipollent to ψ, by (10.11) below, but first notice that since $v \cdot 1_{w^{-1}(a)} = w(v) \cdot 1_a$ one has

$$n \cdot (\lambda\psi + v) = \lambda(n \cdot \psi) + w(v).$$

(10.12) LEMMA. *Given* $u \in U_a - \{1\}$, *then for* $m = m(u)$ *one has*

$$m \cdot \psi = \psi - 2k1_a$$

where $k = \psi_a(u)$.

PROOF: For $b \in \Phi$ and $x \in U_b$ with $\psi_b(x) = \ell$ we shall evaluate $(m \cdot \psi)_b(x)$. Assume first that $b \neq \pm a$, and let Ψ be the set of roots $c \in \Phi$ such that

$1_c = p1_a + q1_b$ with $q > 0$ (see Figure 10.10 in which $\Psi = \{b, c, d\}$).

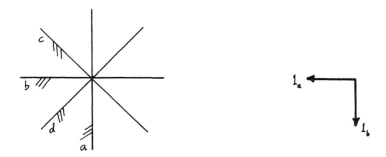

Figure 10.10

Notice that both $\Psi \cup \{a\}$, and $\Psi \cup \{-a\}$ are a full set of positive roots. Also, the reflection r, switching a and $-a$, stabilizes Ψ. Now for $c \in \Psi$, set

$$h(c) = pk + q\ell;$$

in particular $h(b) = \ell$. By (V2) the product

$$\prod_{c \in \Psi} U_{c,n(c)} = U'$$

is a group, and one has

$$U_b \cap U' = U_{b,\ell}$$

and

$$U_{r(b)} \cap U' = U_{r(b),h(r(b))}.$$

Since $1_{r(b)} = 1_b - 2(1_a \cdot 1_b)1_a$, we have $h(r(b)) = \ell - 2(1_a \cdot 1_b)k$. Furthermore if we write $m = vuv'$ where $v, v' \in U_{-a}$, then by (10.10) $\psi_a(u) = k$ implies $\psi_{-a}(v) = \psi_{-a}(v') = -k$. Using (V2) again we see U' is normalized by v, u, and v', and hence by m. Therefore

$$m^{-1}U_{b,\ell}m = m^{-1}U_b m \cap U' = U_{r(b)} \cap U' = U_{r(b),h(r(b))}.$$

Hence

$$(m \cdot \psi)_b(x) = \psi_{r(b)}(m^{-1}xm) = \ell' \geq h(r(b)) = \ell - 2(1_a \cdot 1_b)k.$$

Similarly $\ell \geq \ell' - 2(1_a \cdot 1_{r(b)})k = \ell' + 2(1_a \cdot 1_b)k$, and therefore $\ell' = \ell - 2(1_a \cdot 1_b)k$. Thus $(m \cdot \psi)_b(x) = \psi_b(x) - 2k1_a \cdot 1_b$, as required (for $b \neq \pm a$).

For $b = a$, $1_a \cdot 1_b = 1$, and (V3) immediately gives

$$(m \cdot \psi)_b(x) = \psi_{-b}(m^{-1}xm) = \psi_b(x) - 2k1_a \cdot 1_b.$$

For $b = -a$, then using (V3) and the fact (see (A.1) in Appendix 1) that $m = m(v)$ for the second equality, and $\psi_b(v) = -k$ and $1_a \cdot 1_b = -1$ for the third, one has

$$(m \cdot \psi)_b(x) = \psi_{-b}(m^{-1}xm) = \psi_b(x) - 2\psi_b(v) = \psi_b(x) - 2k1_a \cdot 1_b.$$

This completes the proof that $m \cdot \psi = \psi - 2k1_a$.   □

We now take some given valuation ψ of the root groups, and let A denote the set of valuations equipollent to ψ, namely all $\psi + v$ with $v \in V$. It has the structure of an affine space by taking the distance between $\psi + v$ and $\psi + w$ to be the length of the vector $v - w$. We know that each root $a \in \Phi$ corresponds to a half-space of V, and hence to a half-space of A, namely $(a, 0) = \{\psi + v \mid v \cdot 1_a \geq 0\}$. More generally set $\Gamma_a = \psi_a(U_a - \{1\})$ and for each $k \in \Gamma_a$ define the *affine root* (a, k) as

$$(a, k) = \{\psi + v \mid v \cdot 1_a \geq -k\}$$

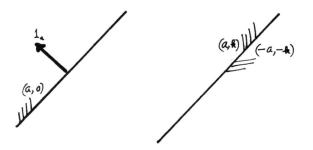

Figure 10.11

In Figure 10.11 $k > 0$. Let Φ^{aff} be the set of affine roots, which we denote by Greek letters $\alpha, \beta, \ldots$; their boundaries $\partial\alpha$ are called the *walls* of A. If α is the affine root (a, k) we let U_α denote $U_{a,k}$. For $n \in N$, $n \cdot \alpha$ means the set of $n \cdot \psi$ for $\psi \in \alpha$, and we have:

(10.13) LEMMA. $nU_\alpha n^{-1} = U_{n\cdot\alpha}$.

PROOF: Let $\alpha = (a, k)$, and let n induce $w \in W(\Phi)$. One has

$$nU_\alpha n^{-1} = \{u \mid n^{-1}un \in U_\alpha \text{ and } \psi_a(n^{-1}un) \geq k\}$$
$$= \{u \in U_{w(a)} \mid (n\cdot\psi)_{w(a)}(u) \geq k\}$$

Let $v = n\cdot\psi - \psi \in V$, so $k \leq (n\cdot\psi)_{w(a)}(u) = \psi_{w(a)}(u) + v\cdot 1_{w(a)}$. Thus $nU_\alpha n^{-1} = U_\beta$ where $\beta = (w(a), k - v\cdot 1_{w(a)})$, and we must show $\beta = n\cdot\alpha$. This can be seen as follows:

$$\beta = \{\varphi \in A \mid (\varphi - \psi)\cdot 1_{w(a)} \geq -(k - v\cdot 1_{w(a)})\}$$
$$= \{\varphi \in A \mid (\varphi - \psi - v)\cdot 1_{w(a)} \geq -k\}$$
$$= \{\varphi \in A \mid (\varphi - n\cdot\psi)\cdot 1_{w(a)} \geq -k\}$$
$$= \{\varphi \in A \mid (n^{-1}\cdot\varphi - \psi)\cdot 1_a \geq -k\}$$
$$= \{\varphi \in A \mid n^{-1}\cdot\varphi \in \alpha\}$$
$$= n\cdot\alpha$$

□

(10.14) COROLLARY. N acts on A as an affine Coxeter group, and the walls subdivide A into a Coxeter complex.

PROOF: If $\alpha = (a, k)$ then the reflection across the wall $\partial\alpha$ sends v to $r(v) - 2k1_a$, where $r \in W(\Phi)$ is the reflection switching a and $-a$. Now if $m \in M_{a,k} = \{m(u) \mid \psi_a(u) = k\}$ then m induces $r \in W(\Phi)$; and using (10.12) we have

$$m\cdot(\psi + v) = m\cdot\psi + r(v) = \psi - 2k1_a + r(v).$$

Thus m acts on A as the reflection across the wall $\partial\alpha$. It therefore only remains to show that N sends walls to walls. This follows from the preceding lemma because if $\alpha = (a, k)$ is an affine root, then $nU_\alpha n^{-1} = U_\beta$ for some affine root β, and we have $n\cdot\alpha = \beta$. □

We let $W(\Phi^{\text{aff}})$, or simply W^{aff}, denote this affine Coxeter group, and let H be the kernel of the action of N on A. Thus $N/H \cong W^{\text{aff}}$.

We can now define the subgroup B of our affine Tits system. Take a chamber c of A (regarded as a Coxeter complex) and let $\Phi^{\text{aff}}(c)$ denote the affine roots containing c. Then B is the group generated by H and the U_α for $\alpha \in \Phi^{\text{aff}}(c)$. Before proving (B, N) is an affine Tits system, we need some technical lemmas.

(10.15) LEMMA. *Let $u \in U_a$ and $v \in U_{-a}$ with $\psi_a(u) + \psi_{-a}(v) > 0$. Then $uv = v_1 h u_1$ where $u_1 \in U_a$, $v_1 \in U_{-a}$ and $h \in H$. Furthermore $\psi_a(u_1) = \psi_a(u)$ and $\psi_{-a}(v_1) = \psi_{-a}(v)$.*

PROOF: If we set $L_a = \langle U_a, U_{-a}, H \rangle$ and $M_a = \langle H, m(u) \mid u \in U_a \rangle$, then it is an exercise having nothing to do with valuations (see Bruhat-Tits [1972] page 108 (6.1.2) (7)) that

$$L_a = M_a U_a \cup U_{-a} H U_a.$$

Moreover $uv \notin M_a U_a$, otherwise for some $u' \in U_a$ we would have $uvu' \in N$, so $m(v) = uvu'$, hence by (10.10) $\psi_a(u) = -\psi_{-a}(v)$, contradicting our hypothesis. Thus $uv = v_1 h u_1$, where $v_1 \in U_{-a}$, $u_1 \in U_a$ and $h \in H$.

For the final statement we may suppose $u, v \neq 1$. Therefore $v_1 \neq 1$ and writing $m = m(v_1)$ we have $v_1 = u_2 m u_3$ where $u_2, u_3 \in U_a$, and by (10.10)

$$\psi_a(u_2) = -\psi_{-a}(v_1)$$

This gives $v = u^{-1} u_2 m u_3 h u_1 = u^{-1} u_2 m h u_3^h u_1 \in U_a N U_a$ Again (10.10) implies

$$\psi_{-a}(v) = -\psi_a(u^{-1} u_2).$$

Therefore $\psi_a(u^{-1} u_2) < \psi_a(u)$, and hence, using (V1)

$$\psi_a(u^{-1} u_2) = \psi_a(u_2).$$

Therefore $\psi_{-a}(v) = -\psi_a(u_2) = \psi_{-a}(v_1)$, and similarly $\psi_a(u) = \psi_a(u_1)$. $\square$

If we select some chamber x of the spherical Coxeter complex $W(\Phi)$ then $\Phi = \Phi^+ \cup \Phi^-$ where the *positive roots* Φ^+ (or *negative roots* Φ^-) are those containing (or not containing) x - see Chapter 6 section 4. We now define:

$$U(c) = \langle U_{a,k} \mid c \in (a, k) \rangle$$
$$U^+(c) = \langle U_{a,k} \mid a \in \Phi^+ \text{ and } c \in (a, k) \rangle$$
$$U^-(c) = \langle U_{a,k} \mid a \in \Phi^- \text{ and } c \in (a, k) \rangle$$

Since H normalizes each $U_{a,k}$, $U(c)$ is normal in B, and $B = U(c) \cdot H$. Furthermore we can describe the structure of $U(c)$ as follows.

(10.16) LEMMA. (i) $U(c) = U^+(c)U^-(c)(N \cap U(c))$, and $N \cap U(c) \subset H$.
(ii) For any $a \in \Phi$, $U_a \cap U(c) = U_\alpha$ where α is minimal with respect to
containing c.

PROOF: Set $X_a = U_\alpha$ where α is minimal with respect to containing c,
as in (ii); thus $U(c)$ is the group generated by the X_a for $a \in \Phi$. We set
$H(c) = H \cap U(c)$ and first show that $U^+(c)U^-(c)H(c) = U(c)$. Since this
product is contained in $U(c)$ it suffices to show it remains stable under left
multiplication by X_a for each $a \in \Phi$. Certainly this is true if $a \in \Phi^+$, so we
need only show that the product is the same regardless of the decomposition
of Φ into positive and negative roots.

If Σ denotes the apartment of which Φ is the set of roots, then Φ^+ is
the set of roots containing some chamber $x \in \Sigma$, so it suffices to show the
product is unchanged when we replace x by an adjacent chamber $y \in \Sigma$.
If $a \in \Phi$ is the root containing x but not y, then y gives positive roots
$\Phi^{+'} = \{-a\} \cup \Phi^+ - \{a\}$, and negative roots $\Phi^{-'} = \{a\} \cup \Phi^- - \{-a\}$. We
let X'_a (or X'_{-a}) be the group generated by the X_b for $b \in \Phi^+$ and $b \neq a$
(or $b \in \Phi^-$ and $b \neq -a$). Then with an obvious notation $U^{+'}(c) = X'_a X_{-a}$
and $U^{-'}(c) = X'_{-a} X_a$.

Notice that X_a and X_{-a} normalize both X'_a and X'_{-a}, and moreover
by (10.15) $X_a X_{-a} H(c) = X_{-a} X_a H(c)$. Therefore:

$$U^+(c)U^-(c)H(c) = X'_a X_a X_{-a} X'_{-a} H(c)$$
$$= X'_a X'_{-a} X_a X_{-a} H(c)$$
$$= X'_a X'_{-a} X_{-a} X_a H(c)$$
$$= X'_a X_{-a} X'_{-a} X_a H(c)$$
$$= U^{+'}(c)U^{-'}(c)H(c).$$

As explained above, this shows that $U(c) = U^+(c)U^-(c)H(c)$.

We now show $H(c) = N \cap U(c)$, so take $n \in N \cap U(c)$ and write
$n = u^+ u^- h$ with an obvious notation. Then $u^- h$ fixes the chamber x' of Σ
opposite x, and u^+ sends it to a chamber opposite x; but $n(x') \in \Sigma$, hence
$n(x') = x'$, and so n acts trivially on Σ. This shows $n \in H$, and completes
the proof of (i).

To prove (ii) let $u \in U_a \cap U(c)$ where without loss of generality $a \in \Phi^+$.
By (i) $u = u^+ u^- h$ with the notation above, and $u^- h$ fixes x', so $u(x') =
u^+(x')$. Since $U^+ = \langle U_a \mid a \in \Phi^+ \rangle$ acts simple-transitively on chambers x'
opposite x (see Chapter 6 Exercise 16 and (6.15)) we have $u = u^+$. Again

using (6.15), after suitably ordering the roots of Φ^+ (according to a gallery determined by the longest word in $W(\Phi)$), the map $\prod\limits_{a\in\Phi^+} U_a \to U^+$ is a bijection. By (V2) this holds equally for $\prod\limits_{a\in\Phi^+} X_a \to U^+(c)$, and therefore $u \in X_a$, completing the proof. $\square$

In the next lemma, which is needed to prove (BN2), $\alpha = (a,k)$ is any affine root, and $-\alpha^+ = (-a,\ell)$ where ℓ is minimal satisfying $k + \ell > 0$. Thus α and $-\alpha^+$ have parallel walls with no wall between - see Figure 10.12.

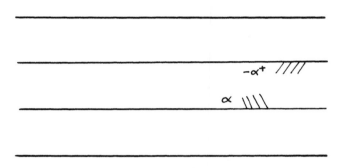

Figure 10.12

(10.17) LEMMA. *If $m \in M_{a,k}$, and L_α denotes the group generated by H, U_α and $U_{-\alpha}$, then $L_\alpha = (U_\alpha m H U_\alpha) \cup (U_\beta H U_\alpha)$, where $\beta = -\alpha^+$.*

PROOF: Let $X_1 = U_\alpha m H U_\alpha$ and $X_2 = U_\beta H U_\alpha$. Since $m \in U_a U_{-a} U_a$ and $U_\beta \subset U_{-\alpha}$ we see that $X_1 \cup X_2 \subset L_\alpha$.

Conversely let $X = X_1 \cup X_2$. Since $m U_\alpha m^{-1} = U_{-\alpha}$ we have $L_\alpha = \langle H, U_\alpha, m \rangle$, and so it suffices to show that $HX \subset X$, $U_\alpha X \subset X$ and $mX \subset X$. The first is immediate because H normalizes U_α and U_β, and m normalizes H. As to the second, $U_\alpha X_1 = X_1$ is clear, and $U_\alpha X_2 \subset X_2$ follows from (10.15).

Finally we show $mX \subset X$. First notice that

$$mX_2 = mU_\beta H U_\alpha \subset mU_{-\alpha} H U_\alpha \subset U_\alpha m H U_\alpha = X_1.$$

Moreover since $m^2 \in H$ we have

$$mX_1 = mU_\alpha m H U_\alpha = U_{-\alpha} H U_\alpha,$$

so it suffices to show that $U_{-\alpha}HU_\alpha \subset X$. To see this let $u \in U_{-\alpha} - U_\beta$, so $\psi_{-a}(u) = -k$ and there exist $v, v' \in U_\alpha$ with $vuv' \in M_{a,k} \subset mH$. Therefore $u \in U_\alpha m H U_\alpha$, and hence

$$U_{-\alpha} \subset U_\beta \cup U_\alpha m H U_\alpha.$$

Therefore $U_{-\alpha}HU_\alpha \subset X_2 \cup X_1 = X$, completing the proof. □

We are now ready to prove the main theorem of this section.

(10.18) THEOREM. *If G denotes the group generated by the U_a, then (B,N) is an affine Tits system for G.*

PROOF: We verify axioms (BN0) - (BN3); recall that c is a chamber of A, and B is generated by H and $U(c)$.

(BN0) Certainly B and N are subgroups of G, and for any $u \in U_a$ with $\psi_a(u) = k$, either $c \in (a,k)$ in which case $u \in B$, or $c \in (-a, -k)$ in which case $m(u)um(u) \in B$ and $u \in \langle B, N \rangle$. Thus $G = \langle B, N \rangle$.

(BN1) From the action of N on A we know $N/H \cong W^{\mathrm{aff}}$; moreover $H \subset B \cap N$, so we need only show $B \cap N \subset H$. As mentioned earlier H normalizes $U(c)$, so $B = U(c)H$ and since $N \cap U(c) \subset H$ by (10.16)(i) we have $N \cap B \subset H$.

(BN2) Let s be a reflection in some wall $\partial\alpha$ containing a panel of c; we must show $BsBwB \subset BswB \cup BwB$ for any $w \in W^{\mathrm{aff}}$. Let $c \in \alpha$, and set $\Psi = \Phi^{\mathrm{aff}}(c) - \{\alpha\}$. If U' denotes the group generated by the U_β for $\beta \in \Psi$, then from (10.16)(i) we have

$$B = U'HU_\alpha.$$

Moreover s stabilizes Ψ, and hence normalizes U', so this gives

$$sB \subset BsU_\alpha. \tag{*}$$

Recall that L_α is a group containing s (or rather its inverse image in N), and therefore $L_\alpha wB = L_\alpha swB$. Replacing w by sw if necessary we may assume $w^{-1}(\alpha)$ contains c, in which case $w^{-1}U_\alpha w \subset B$. Now letting β denote the translate of $-\alpha$ which is minimal with respect to containing c, (10.17) gives

$$L_\alpha wB = U_\alpha s H U_\alpha wB \cup U_\beta H U_\alpha wB \subset BswB \cup BwB$$

Using (*) this implies

$$BsBwB \subset BsU_\alpha wB \subset BL_\alpha wB \subset BswB \cup BwB$$

as required.

(BN3) With the notation above, $sU_\alpha s = U_{-\alpha}$, and by (10.16)(ii) $U_{-\alpha} \not\subset B$. Therefore $sBs \neq B$. $\quad\square$

If we let Δ denote the affine building determined by B and N, then we obtain an apartment system $\mathcal{A}$ as follows. Treat chambers of Δ as left cosets gB as in Chapter 5, and let Σ be the apartment whose chambers are all nB, for $n \in N$. The images of Σ under G are the apartments of $\mathcal{A}$; they correspond to the apartments of the spherical building for $(U_a)_{a \in \Phi}$ (because both correspond to the conjugates of the set $(U_a)_{a \in \Phi}$), and hence $(\Delta, \mathcal{A})^\infty$ is this spherical building.

Moreover there is a canonical isomorphism between Σ and the simplicial complex of A, given by $nB \leftrightarrow w(c)$ where $w = nH \in W^{\mathrm{aff}}$. Let s be the point of Σ corresponding to $\psi \in A$. If we take the metric on Σ induced by A, and let φ be the valuation determined by Σ and s, as in section 3, then $\varphi = \psi$. Indeed if $u \in U_a$ with $\psi_a(u) = k$, then $m(u) \in N$ fixes a wall of A at distance k from ψ (in the $+a$ to $-a$ direction), and therefore a wall of Σ, similarly at distance k from s. This must be the boundary wall of $\Sigma \cap u\Sigma$, and therefore $\varphi_a(u) = k$. We have therefore proved:

(10.19) COROLLARY. *Any set of root data with a discrete valuation arises from an apartment system in an affine building.* $\quad\square$

5. The Classification.

This section deals mainly with the classification of affine buildings $(\Delta, \mathcal{A})$ having rank ≥ 4 (and a connected diagram). In this case $(\Delta, \mathcal{A})^\infty$ has rank ≥ 3, so we can apply the classification of spherical buildings in Chapter 8.

The first step (10.20) is to show that when $(\Delta, \mathcal{A})^\infty$ is Moufang (in particular when its rank is at least 3), the problem reduces to classifying root data with valuation. If $(\Delta, \mathcal{A})^\infty$ has rank 2, then it is a generalised m-gon which might not be Moufang, and a classification is not possible, as we shall explain.

(10.20) THEOREM. *The $(\Delta, \mathcal{A})$ for which $(\Delta, \mathcal{A})^\infty$ is Moufang correspond to equivalence classes of root data with valuation.*

PROOF: By (10.19) every set of root data with valuation arises from a suitable affine system $(\Delta, \mathcal{A})$, uniquely determined by (10.5). Conversely we saw in section 3 how a given $(\Delta, \mathcal{A})$ gives rise to a set of root data with valuation when $(\Delta, \mathcal{A})^{\infty}$ is Moufang. That involved choosing a point s in an apartment A and assigning a Euclidean metric to A. The choice of apartment is irrelevant because if A' is any other apartment there is an automorphism of $(\Delta, \mathcal{A})$ inducing an isometry from A to A'; and the choice of metric and point s gives an equivalent evaluation, as explained at the end of section 3.   □

We now show that a valution $\psi = (\psi_a)$ is determined up to equipollence by a single ψ_a.

(10.21) THEOREM. *If φ and ψ are valuations of the same root groups, then $\varphi_a = \psi_a$ for some $a \in \Phi$ if and only if $\varphi = \psi + v$ for some $v \in V$ perpendicular to 1_a.*

PROOF: Given any root $b \in \Phi$, let $c = r_b(a)$ where r_b is the reflection interchanging b and $-b$, and let $m = m(x)$ for some $x \in U_b - \{1\}$. Given $y \in U_c - \{1\}$, recall that $(m \cdot \psi)_c(y) = \psi_a(m^{-1}ym)$ by definition, and therefore by (10.12) we have

$$\psi_a(m^{-1}ym) = (\psi - 2\psi_b(x)1_b)_c(y)$$
$$= \psi_c(y) - 2\psi_b(x)1_b \cdot 1_c$$
$$= \psi_c(y) + 2\psi_b(x)1_a \cdot 1_b$$

We rewrite this as:

$$2(1_a \cdot 1_b)\psi_b(x) = \psi_a(m^{-1}ym) - \psi_c(y).$$

If $1_a \cdot 1_b \neq 0$, this shows that ψ_a determines ψ_b up to an additive constant. Moreover, if ψ_a and ψ_b are known, so is ψ_c, and therefore ψ is uniquely determined by its components at a set of fundamental roots $a_1, \ldots, a_\ell$; we choose these so that $a_1 = a$.

Let $\varphi_{a_i} = \psi_{a_i} + k_i$ where $k_i \in \mathbf{R}$ is a constant (and of course $k_1 = 0$), and let $v \in V$ be the unique vector such that $v \cdot 1_{a_i} = k_i$. Then the valuation $\psi' = \psi + v$ has the property that $\psi'_{a_i} = \psi_{a_i} + k_i$, and hence $\psi' = \varphi$. Thus $\varphi_a = \psi_a$ implies $\varphi = \psi + v$ where $v \cdot 1_a = 0$, and the converse is a triviality.   □

Single Bond Diagrams of Rank ≥ 4. The single bond cases of rank ≥ 4 are $\widetilde{X}_n = \widetilde{A}_n$ $(n \geq 3), \widetilde{D}_n(n \geq 4)$ or $\widetilde{E}_n(n = 6, 7$ or $8)$, and the classification of spherical buildings (Chapter 8) shows that $(\Delta, \mathcal{A})^{\infty}$ must be the $X_n(K)$ building for some field K (necessarily commutative except in the A_n case). Furthermore if π is any panel in a wall m of $(\Delta,\mathcal{A})^\infty$, then $St(\pi)$ and $St(m)$ both correspond to the projective line $K \cup \{\infty\}$.

Now let Φ denote the roots in some given apartment of $(\Delta,\mathcal{A})^\infty$, and let $a \in \Phi$, and $m = \partial a$. Without loss of generality we identify $St(m)$ with $K \cup \{\infty\}$ so that a corresponds to ∞, and $-a$ to $0 \in K$. The root group U_a is the group of affine translations, and we label its elements by subscripts belonging to K, so that $u_0 = id.$, and $u_x(-a) \in St(m)$ corresponds to the point $x \in K$ of the affine line. By Example 3 in section 2, $\omega_m(a, -a; u_1(-a), u_x(-a)) = v(x)$ for some discrete valuation v of K.

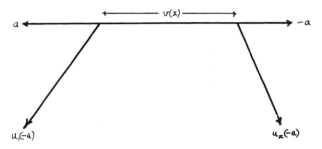

Figure 10.13

It is clear from Figure 10.13 that if $\varphi_a(u_1) = t \in \mathbf{R}$, then $\varphi_a(u_x) = t+v(x)$; in particular after identifying K with U_a, v determines φ_a up to an additive constant. Recall that after multiplying v by a suitable positive real number, we have $v(K^\times) = \mathbf{Z}$, in which case v is called *normalized*.

(10.22) THEOREM. *There is a bijective correspondence between thick affine systems* $(\Delta,\mathcal{A})$ *of type* $\widetilde{X}_n = \widetilde{A}_n, \widetilde{D}_n, \widetilde{E}_6, \widetilde{E}_7$ *or* $\widetilde{E}_8$, *and pairs* (K,v) *where* K *is a field (necessarily commutative except for* $\widetilde{A}_n$) *with a discrete, normalized valuation* v.

PROOF: As mentioned above, $(\Delta,\mathcal{A})^\infty$ is the $X_n(K)$ building, and by (10.20), $(\Delta,\mathcal{A})$ corresponds to an equivalence class of $X_n(K)$ root data. By the discussion above, this gives a discrete, normalized valuation v of K, and although v only determines φ_a up to an additive constant, (10.21) shows that it determines the root data up to equipollence. It therefore only remains to show that *any* v gives a set of root data with valuation.

To see this, let G be the group generated by the U_a, and take a nontrivial representation of G over the field K. Each U_a is then represented as a group of unipotent matrices with a single non-zero entry off the diagonal. If $u \in U_a$ let e_u denote this non-zero element, and set

$$\varphi_a(u) = v(e_u).$$

It remains to check that the (φ_a) satisfy (V0) - (V3) of section 3. In fact (V0) and (V1) are immediate, and to check (V2) and (V3) we first note that if $a \neq \pm b$ then a and b span either an A_2 or an $A_1 \times A_1$ subsystem, because each rank 2 residue is of this type. Thus it suffices to check (V2) and (V3) in an $A_1 \times A_1$ or A_2 system, and this is completely straightforward; we leave it to the reader, and refer to Example 4 of section 3. $\square$

Double Bond Diagrams of Rank ≥ 4. These cases are $\widetilde{F}_4$, $\widetilde{B}_n$ and $\widetilde{C}_n$ for $n \geq 3$. In all such cases $(\Delta, \mathcal{A})^{\infty}$ has two types of walls m, and for at least one type $St(m)$ is a projective line over a field K (not necessarily commutative). Indeed the classification of Chapter 8 shows that a C_2 residue of $(\Delta, \mathcal{A})^{\infty}$ must be a Moufang quadrangle of classical type; that is to say, it arises from those 1-spaces and 2-spaces in a vector space over K, which are totally singular or isotropic under a suitable form of Witt index 2. This implies there are panels π for which $St(\pi)$ corresponds to the 1-spaces in a 2-space over K, hence our assertion above about $St(m)$. The discussion in the single bond case then gives:

(10.13) PROPOSITION. *Each system $(\Delta, \mathcal{A})$ of type $\widetilde{F}_4$, $\widetilde{B}_n$ or $\widetilde{C}_n$, for $n \geq 3$, gives rise to a field K having a discrete valuation v.* $\square$

Remark. In general K is not uniquely determined, because the quadrangle (C_2 residue) and its dual may both arise from a form of Witt index 2 as above, but over different fields (see Chapter 8 section 5). However if $(\Delta, \mathcal{A})$ has type $\widetilde{B}_n$ or $\widetilde{C}_n$, then $(\Delta, \mathcal{A})^{\infty}$ has type C_n and there is a canonical choice for K: if the A_2 residues are Desarguesian (which is always the case for $n \geq 4$), let K be the associated field, and in the C_3 Cayley plane case there is only one choice for K anyway (corresponding to the end node of the double bond - see Chapter 8 section 5).

We come now to the following question. Given a field K with a discrete valuation v, and given a C_n or F_4 building associated to K as above, does there exist an appropriate affine system $(\Delta, \mathcal{A})$? In other words can one find a valuation (φ_a) of the root data inducing the valuation v on K? In general the answer is no, because a C_2 residue (or indeed the whole building, except in the F_4 or C_3 Cayley plane cases) arises from a (σ, ϵ)-hermitian form and v must certainly be invariant under σ. However, assuming this to be the case, the answer is yes when K is complete with respect to v (at least in the C_n or F_4 case considered here) - see Tits [1986a] p.173.

In general let $G(K)$ be the group generated by the root groups of the C_n or F_4 building, and let $G(\widehat{K})$ be obtained by completing the field with respect to v (considering $G(K)$ as a group of matrices over K, satisfying certain polynomial conditions, $G(\widehat{K})$ is obtained under the same conditions but by extending the field from K to $\widehat{K}$). Then assuming v is invariant under σ as above, one has:

A valuation (φ_a) of the root data inducing v on K exists if and only if $G(K)$ and $G(\widehat{K})$ have the same rank.

In fact if S_K and $S_{\widehat{K}}$ are the (spherical) buildings for $G(K)$ and $G(\widehat{K})$, then by the remarks in the preceding paragraph, $S_{\widehat{K}} = \Delta^\infty$ for some affine building Δ. If $G(K)$ and $G(\widehat{K})$ have the same rank then S_K is a sub-building of $S_{\widehat{K}}$ of the same type and is therefore the union of apartments in some subset $\mathcal{A}$ of all apartments for $S_{\widehat{K}}$; thus $S_K = (\Delta, \mathcal{A})^\infty$. Conversely if $S_K = (\Delta, \mathcal{A})$, then $G(K)$ is generated by the root groups of $(\Delta, \mathcal{A})^\infty$, and so $G(\widehat{K})$ is generated by the root groups of Δ^∞; thus $G(K)$ and $G(\widehat{K})$ have the same rank.

If $G(K)$ is a classical group arising from a (σ, ϵ)-hermitian form, then to say that $G(\widehat{K})$ has the same rank as $G(K)$ means that the Witt index of this form does not increase when we extend the scalars from K to $\widehat{K}$. In the non-classical case we have either a C_3 building with Cayley planes, or an F_4 building, and in the F_4 case the rank can only increase if the building involves Cayley planes or quaternion planes (see the diagrams in Chapter 8 section 5). To say that the rank remains the same amounts to saying that the appropriate Cayley or quaternion division algebra does not split when we pass from K to $\widehat{K}$ (if it does, then $S_{\widehat{K}}$ has type E_7, E_8 or E_7 respectively in the three cases). We conclude this discussion with a theorem.

(10.24) THEOREM. Suppose we have a C_n or F_4 building S over a field K having a discrete valuation v, invariant under σ, as explained earlier. Then v determines an affine system $(\Delta, \mathcal{A})$ with $(\Delta, \mathcal{A})^\infty = S$ if and only if one of the following holds:

(i) S is "classical", arising from a (σ, ϵ)-hermitian form whose Witt index remains the same over $\widehat{K}$.

(ii) S is of type C_3 having Cayley planes, or of type F_4 having Cayley planes or quaternion planes, and the relevant Cayley or quaternion division algebra over K does not split when K is extended to $\widehat{K}$. □

Example. Let $K = \mathbf{Q}$ (the rational numbers). If S involves Cayley planes, then no discrete valuation v gives an affine system $(\Delta, \mathcal{A})$, because there is no Cayley division algebra over the p-adics $\mathbf{Q}_p$. The same is true if S arises from a quadratic form of Witt index n in at least $2n + 5$ variables, because there is no such form over $\mathbf{Q}_p$ (any quadratic form in 5 variables over $\mathbf{Q}_p$ has non-trivial singular vectors).

Finally we state a corollary of the results above, made possible by the fact that if K is complete and has a finite residue field k then K is either a p-adic field (finite algebraic extension of $\mathbf{Q}_p$) or a power series field $k((t))$; in each such case the discrete valuation is unique up to multiplication by a positive real number.

(10.25) COROLLARY. *The thick locally finite affine buildings of rank $n \geq 4$ (with a connected diagram) are the affine buildings of simple algebraic groups of rank $(n-1)$ over a p-adic field or a power series field.* □

The Rank 3 Case. If $(\Delta, \mathcal{A})$ has rank 3 then $(\Delta, \mathcal{A})^\infty$ is a generalized m-gon, for $m = 3, 4$ or 6. A classification is impossible because of a general construction (Ronan [1986]) in which one starts with a single chamber and builds outwards: for each rank 2 residue there is complete freedom of choice amongst all rank 2 buildings having the appropriate type and parameters (number of chambers per panel). The building at infinity, however, can not be chosen arbitrarily; for example it cannot be finite! In the $\widetilde{A}_2$ case, Van Maldeghem [1987] and [1988] shows that the projective plane at infinity is coordinatized by a ternary ring having a discrete valuation (in a sense made precise in those papers), and any such plane arises as a building at infinity of an $\widetilde{A}_2$ building.

6. An Application.

In this section we apply the classification of affine buildings, and the results of Chapter 4, to obtain a result in finite group theory, following recent work of Kantor, Liebler and Tits [1987]. To simplify the exposition we deal with only one case of their work, namely $\widetilde{D}_4$, though a similar argument works for other affine diagrams of rank at least 4.

Consider a finite group G acting transitively on a chamber system C

of type $\widetilde{D}_4$:

We assume the D_4 residues are buildings (though in view of transitivity one can assume less, such as A_2 residues being Desarguesian planes); in particular these residues are $D_4(k)$ buildings where $k = \mathbf{F}_q$ is a finite field of characteristic p.

(10.26) THEOREM. *Except for the $q = 2$ case, no such group G can exist.*

Remark. When $q = 2$ a family of examples was constructed by Kantor [1985]; see also Tits [1986b] (3.2).

PROOF: By Theorem 4.9 the universal cover $\widetilde{C}$ is a building of type $\widetilde{D}_4$, and hence by the classification of section 5 it is the $D_4(K, v)$ building, where K is a commutative field with a discrete valuation v and residue field k. Without loss of generality we may take K to be complete in which case it is either a finite extension of $\mathbf{Q}_p$ (when char $K = 0$), or the power series field $k((t))$ (when char $K = p$).

Now consider the group G. By Chapter 4 Exercise 8, G lifts to a group $\widetilde{G}$ acting transitively on $\widetilde{C}$; moreover the stabilizer in $\widetilde{G}$ of a vertex of $\widetilde{C}$ is isomorphic to the stabilizer in G of its image in C. In particular vertex stabilizers in $\widetilde{G}$ are finite, and in fact this is the starting point for the Kantor-Liebler-Tits paper [loc. cit.].

The argument is now roughly as follows (see below for more details): if x is a vertex of type $i = 1, 2, 3$ or 4 then the finite simple orthogonal group $O_8^+(q)$ acts on $St(x)$, and acquires a non-trivial projective representation in the 8-dimensional K-vector space V, for $D_4(K)$. If char $K = 0$ this forces $q = 2$ (by a result in representation theory which we shall simply "pull out of a hat"). On the other hand if char $K = p$ then this is the natural representation, but we can play off the actions of four separate $O_8^+(q)$ (for vertices of types 1, 2, 3 and 4) to show their representations cannot coexist in V.

To fill in the details of this argument, a theorem of Seitz [1973] shows that, since the stabilizer $\widetilde{G}_x$ is transitive on $St(x)$, it contains a subgroup inducing the simple orthogonal group $O_8^+(q)$ on $St(x)$; let Γ_i be the smallest such subgroup. Since the full automorphism group of $\widetilde{C}$ contains an orthogonal group $O_8(K)$ as a normal subgroup with a solvable quotient, the simplicity of $O_8^+(q)$ implies that Γ_i lies in $O_8(K)$. Thus Γ_i has a non-trivial projective representation in 8-dimensions over K. We now consider separately the cases where the characteristic of K is 0 or p.

Case 1. char $K = 0$. In this case the smallest dimension for a non-trivial projective representation of $O_8^+(q)$ is known (see Landazuri-Seitz [1974]), and it is greater than 8 in all cases except $q = 2$ for which an 8-dimensional representation exists. Furthermore a theorem of Feit and Tits [1978] then implies that Γ_i itself can have no characteristic zero, projective representation in dimension ≤ 8 if $q \neq 2$. Thus if char $K = 0$, then $q = 2$. We shall now complete the proof by showing char $K = p$ is impossible.

Case 2. char $K = p$. By Exercise 9, regardless of the characteristic, the stabilizer of the vertex x in $O_8(K)$ is $O_8(\mathcal{O})$ where $\mathcal{O}$ is the valuation ring of K. The elements of $O_8(\mathcal{O})$ congruent to the identity modulo the maximal ideal of $\mathcal{O}$ form a normal subgroup which is a pro-p-group, and its quotient is an 8-dimensional orthogonal group over k. Thus the finite subgroup Γ_i has a normal p-subgroup U_i whose quotient G_i is an 8-dimensional orthogonal group with $G_i/Z(G_i) = O_8^+(q)$. We shall argue that $U_i = 1$.

Since we are in the characteristic p case, U_i is a unipotent subgroup (i.e. can be put in upper triangular form with diagonal entries all 1), and hence acts trivially on a totally singular 1-space V_1. However Γ_i cannot fix V_1 otherwise it would fix a sector-face having vertex x and direction V_1 in the building at infinity, contradicting the fact that it acts transitively on $St(x)$. Therefore under the action of Γ_i, V_1 generates a non-trivial module V_0 for the orthogonal group G_i. No such module exists in dimension < 8; so $V_0 = V$, and since U_i acts trivially on V_0, we have $U_i = 1$.

We can now complete the argument by comparing the actions of $\Gamma_1, \ldots, \Gamma_4$ on V, to obtain a contradiction. We have established that Γ_i is an orthogonal group; and V is a natural module for Γ_i, when $i = 1, 2, 3$ or 4. Let B be the stabilizer in $\widetilde{G}$ of a chamber c, and let P_j be the stabilizer of the j-panel of c (e.g. $\Gamma_1 = \langle P_0, P_2, P_3, P_4 \rangle$). Let U be the normal Sylow-p-subgroup of B; it fixes a 1-space $X \subset V$, and under the action of P_j generates either a 1-space or a 2-space. In fact considering V as a natural Γ_1-module, X is a totally singular 1-space, and its stabilizer is, without loss of generality, $\langle P_0, P_2, P_3 \rangle$; also it lies in a t.s. 2-space whose stabilizer is $\langle P_2, P_3, P_4 \rangle$. Thus under P_4, X generates a 2-space, and under P_0, P_2, P_3 a 1-space.

A similar situation holds for Γ_2, Γ_3 and Γ_4, so each node j of the diagram must be assigned a 1 or a 2, such that in *each* D_4 subdiagram exactly one end node has a 2, and the others have a 1.

This is plainly impossible. Therefore these representations cannot coexist in V, and the characteristic p case cannot occur.   □

Remark. In the preceding theorem we have dealt with the generic case. When $q = 2$, Kantor, Liebler and Tits are able to show that Δ has to be over $\mathbf{Q}_2$ (rather than an extension thereof) and that there is essentially only one possibility for the group $\widetilde{G}$ acting on Δ.

Notes. Sections 1, 2, 3 and 5 of this chapter are adapted from Tits [1986a], where the non-discrete case is also dealt with (see Appendix 3), and Section 4 is extracted from Chapter 6 of Bruhat-Tits [1972]. For the classification of locally finite affine buildings (10.25), see Tits [1979].

Exercises to Chapter 10

In all these exercises K is a field with a discrete valuation v, and k denotes its residue field.

1. Let T be the tree $\widetilde{A}_1(K, v)$ of Example 2. Show that $\mathcal{A}(K)$ (Example 1) is the set of all possible apartments if and only if K is complete with respect to v. [HINT: Let z be any end of T, and let $x, y_1, y_2, \ldots$ be ends of apartments in $\mathcal{A}(K)$ such that $[x, y_i] \cap [x, z] \subsetneq [x, y_{i+1}] \cap [x, z]$. With a suitable choice of basis, an element of $SL_2(K)$ sending $[x, y_1]$ to $[x, y_i]$ has the form $\left(\begin{smallmatrix} 1 & a_i \\ 0 & 1 \end{smallmatrix}\right)$ - consider the sequence $a_1, a_2, \ldots$].

2. Consider the building $\widetilde{A}_n(K, v)$ and show $\mathcal{A}(K)$ is the set of all possible apartments if and only if K is complete with respect to v. [HINT: Use Exercise 1 and the trees $T(\pi)$ in section 2].

3. If K is a field, identify the projective line $K \cup \{\infty\}$ with the 1-spaces in a 2-space, by setting $x \mapsto \left(\begin{smallmatrix} x \\ 1 \end{smallmatrix}\right)$ and $\infty \mapsto \left(\begin{smallmatrix} 1 \\ 0 \end{smallmatrix}\right)$. The affine transformation $x \mapsto ax + b$ can then be represented by the matrix $\left(\begin{smallmatrix} a & b \\ 0 & 1 \end{smallmatrix}\right)$. If the projective valuation ω of section 1 is invariant under the affine group

(Example 3), show it is invariant under the projective group. [HINT: It suffices to consider invariance under $\left(\begin{smallmatrix} 0 & 1 \\ -1 & 0 \end{smallmatrix}\right)$, which for $a, b, c, d \in K^\times$ means $\omega(a, b; c, d) = \omega(-a^{-1}, -b^{-1}; -c^{-1}, -d^{-1})$].

4. Let T be a (discrete) tree, and v some fixed vertex. For any vertex $x \neq v$, let $n = n(v, x)$ denote the number of edges from v to x, and define $d(v, x) = 1 - 2^{-n}$. This distance d determines a metric on T, such that the projective valuation on T^∞ is not discrete.

5. If a sector-face X contains a sector-face X_1 of the same dimension, show that X_1 is a translate of X (in any apartment containing X), and therefore parallel to X. Conclude that for sector-panels the property of being asymptotic is an equivalence relation, finer than that of being parallel.

6. Let A_1, A_2, A_3 be apartments of an affine building, such that $A_i \cap A_j$ is a half-apartment, for each $i, j \in \{1, 2, 3\}$. Show that $A_1 \cap A_2 \cap A_3$ is non-empty (though it might not contain any chambers). [HINT: First consider trees, then use $T(m)$].

7. If ψ is a valuation of root groups (satisfying (V0) - (V3)), show that $\lambda\psi + v$ is too.

8. A valuation of root groups is called *special* if $0 \in \Gamma_a$ for each $a \in \Phi$ [recall $\Gamma_a = \varphi_a(U_a - \{1\})$]. Show that the special valuations obtained from an affine building, using a point $s \in A$ as in section 3, are precisely those for which s is a special vertex.

9. Let Δ be the affine building with a single bond diagram $\widetilde{X}_n$ as in section 5, obtained using a field K with discrete valuation v, and let G be generated by the root groups. Regarding G as a matrix group, as in section 5, show that the subgroup stabilizing a special vertex is obtained by taking all those matrices with entries in the valuation ring $\mathcal{O}$ (given a suitable choice of basis, of course).

10. Let X be an $\widetilde{A}_1$ building (a tree). For a vertex x and chamber c on x, let $U_{x,c}$ be the set of ends $e = S^\infty$ where S has vertex x and base chamber c; call this a *basic open set*.

    (i) Show that the intersection of two basic open sets is a union of basic open sets.

    (ii) The $U_{x,c}$ are a basis for a Hausdorff topology on X^∞; if each vertex has valency $\leq s$ for some finite number s, show that this topology makes X^∞ compact.

    (iii) Show that $(X, \mathcal{A})^\infty$ is not compact if $\mathcal{A}$ is not complete (assuming X^∞ is an infinite set).

Chapter 11
TWIN BUILDINGS

This chapter will lay down the basics of twin buildings, with references to the literature. Twin buildings are a generalisation of spherical buildings, and arose initially from Kac-Moody groups.

1. Twin Buildings and Kac-Moody Groups.

When dealing with spherical buildings in Chapter 6, root groups played an important role, and were used in defining Moufang buildings. Such buildings, when spherical, arise naturally from groups of Lie type. When non-spherical, they arise from generalisations of these groups, called Kac-Moody groups, which are infinite dimensional, but of finite rank. In terms of buildings, the appropriate generalisation is from spherical buildings to twin buildings, and although the groups are infinite dimensional, the buildings retain finite rank, and are therefore finite dimensional.

When this book was first published twin buildings had not yet been defined, though the group-theoretic ingredients had already been published by Tits [1987] in his paper on Kac-Moody groups. There he defines a Kac-Moody group G as generated by a set of root groups U_α parameterized by the roots α of a Coxeter complex W. The set of roots can be split (in many different ways) into two parts: a positive part and a negative part. The positive root groups generate a group U_+ and the negative root groups generate a group U_-. Let N denote the subgroup of G normalizing the set of root groups, and let T be the subgroup normalizing each individual root group. As with groups of Lie type, N/T is isomorphic to W. Let B_+ and B_- denote the groups generated by T along with U_+ and U_- respectively—the pairs (B_+,N) and (B_-,N) are BN-pairs for G, as in Chapter 5.

These BN-pairs generate two buildings Δ_+ and Δ_-, whose chambers can be regarded as the left cosets in G of B_+ and B_- respectively. As in Chapter 5, the usual Bruhat decompositions of G, namely $G = B_+WB_+$ and $G = B_-WB_-$, give the W-valued distance between chambers in the same building. A further decomposition B_+WB_- gives a W-valued "codistance" between chambers of Δ_+ and Δ_-. This provides (Δ_+,Δ_-) with the structure of a twin building, which we define below. An example of a twin building in the affine case, for the group $\mathrm{SL}_n(k[t,t^{-1}])$, is given in Section 3.

When W is finite (the spherical case) it has a unique longest word, which conjugates B_+ to B_-. This allows an identification of Δ_+ with Δ_-, and each spherical building can be naturally twinned with itself, as we shall see later. When W is infinite there is no longest word, and the subgroups B_+ and B_- are not conjugate in G. This implies that while B_+ stabilises a chamber in Δ_+, it acts quite differently on Δ_-, an action that has been used over finite ground fields to study finiteness properties (of groups such as $\mathrm{SL}_n(k[t])$ by Abramenko [1996]), and to construct interesting new classes of lattices (Rémy [1999], Carbone and Garland [2003], Rémy and Ronan [2006]). Twin buildings are, however, similar to spherical buildings in the sense that a chamber in one building of the twin can be opposite a chamber in the other. They can be seen as a generalisation of spherical buildings, and can be used to prove results for Kac-Moody groups, just as spherical buildings are used for groups of Lie type. For example, Abramenko and Mühlherr [1997] have used them in obtaining a Curtis-Tits presentation when the rank 2 residues are spherical, Caprace and Mühlherr [2006] have used them in proving isomorphism theorems, and Caprace and Rémy [2006] have used them over finite ground fields in proving a simplicity theorem.

Before giving the definition of a twin building, it is worth rephrasing the definition of a building, following Tits [1992]. Earlier in this book, our canonical generators for W were called r_i; in order to simplify notation, we shall now use the symbol s to mean r_i, without specifying i, and refer to i-adjacent chambers as being s-*adjacent*. The length of an element w in W will now be written $|w|$ (rather than $\ell(w)$ as we wrote earlier), and the W-distance from a chamber x to a chamber y will be written $w(x,y)$, rather than $\delta(x,y)$. With these small amendments to the notation, the definition of a building in terms of the W-distance can be given as:

(B1) $w(x,y) = 1$ *if and only if $x = y$.*

(B2) *If $w(x,y) = w$ and if y is s-adjacent to z, then $w(x,z) = w$ or ws.*

(B3) *Each s-panel π is a face of at least two chambers, and for any chamber* *x, exactly one of these chambers on π is at minimal distance from x (i.e. for all other chambers z on π, w(x,z) = ws where |ws| = |w|+1).*

Exercise 1.   Using the definition of a building given on page 27, show that (B1) to (B3) are satisfied. Conversely show that if (B1) to (B3) hold then the axiomatic condition on page 27 is satisfied. In other words the definition above is equivalent to the definition given earlier in Chapter 3.

Note that if x is a chamber, and $π$ an s-panel, as in (B2) and (B3), then $\text{proj}_π x$ is the unique chamber z on $π$ for which $w(x,z)$ is the shorter of w and ws.

A *twin building Δ of type W* is a pair of buildings $(Δ_+, Δ_-)$, both of type W, along with a *codistance* function w^* defined below. The buildings $Δ_+$ and $Δ_-$ are the *components* of $Δ$, and for chambers x and y in the same component, the W-distance $w(x,y)$ satisfies the axioms above. When x and y are in different components, the axioms for the codistance $w^*(x,y)$ are the following.

(TB1) $w^*(y,x) = w^*(x,y)^{-1}$.

(TB2)   *If $w^*(x,y) = w$ and if y is s-adjacent to z, then $w^*(x,z) = w$ or ws.*

(TB3)   *Given a chamber x in one component of $Δ$, and an s-panel $π$ in the other component, then exactly one chamber on $π$ is at maximal codistance from x (i.e. for all other chambers z on $π$, $w^*(x,z) = ws$ where $|ws| = |w| - 1$).*

If z is adjacent to x, instead of y in (TB2), then by (TB1) and (TB2) $w^*(z,y) = w$ or sw.

Some notation.   In the setting of twin buildings, an *isometry* will mean a map preserving distances and codistances. An element w in W that can be written in reduced form $s_1 \ldots s_n$ with $s = s_1$ is called s-*extended* on the left (and if with $s = s_n$ is called s-*extended* on the right). Otherwise we call it s-*reduced* on the left (or on the right). Notice that w is s-extended on the left (or on the right) if and only if $|sw| < |w|$ (or $|ws| < |w|$).

The Spherical Case.   Let W be a finite Coxeter group, and w_0 its longest word. A spherical building $Δ$ of type W is canonically twinned with itself, as follows. Let $Δ_+$ and $Δ_-$ be copies of $Δ$, and for any chamber c in $Δ$ let c_+ and c_- be its copies in $Δ_+$ and $Δ_-$ respectively. Define the distance functions on $Δ_+$ and $Δ_-$ by $w(x_+,y_+) = w(x,y)$, and $w(x_-,y_-) = w_0 w(x,y) w_0$; then define a codistance on (Δ_+, Δ_-) by $w^*(x_+, y_-) = w(x,y)w_0$, and $w^*(x_-, y_+) = w_0 w(x,y)$. It is straightforward to check that this makes (Δ_+, Δ_-) a twin building.

Opposites. Recall that in a spherical building we call two chambers x and y opposite if the W-distance $w(x,y) = w_0$; in terms of the canonical twinning above this means that $w^*(x_+, y_-) = 1$. When W is infinite there is no longest word, but the concept of opposite chambers applies to all twin buildings: two chambers x and y, one in one component and one in the other, are called *opposite* if $w^*(x,y) = 1$ (and two residues are called *opposite* if they have the same type and contain chambers that are opposite one another).

The definition of a twin building in terms of a W-codistance is a good one to work with, but not always an easy one to verify in a particular case. It may for example be relatively easy to define when two chambers are opposite, but more problematic to ascertain the codistance between two arbitrary chambers, even though this codistance is uniquely determined by knowing all pairs of opposites. With this in mind, Mühlherr [1998] has shown that an opposition relation between the chambers of two buildings yields a twinning if and only if it does so in all pairs of opposite rank 2 residues.

2. Twin Trees.

The simplest case of a non-spherical twin building is a twin tree. These interesting objects are analogous to the generalised polygons that we introduced immediately after the definition of a building, but defined in terms of a distance between vertices. We will do the same here with twin trees, defining them in terms of a codistance between vertices, rather than a W-codistance between edges.

Consider first a single tree; if x and y are distinct vertices there is a unique neighbour of y at lower distance from x. In a twin tree, if x and y are in distinct trees and not opposite one another then there is a unique neighbour of y at higher codistance from x. Here is the definition.

A twin tree is a pair of trees together with a non-negative integer $\mathrm{codist}(x,y) = \mathrm{codist}(y,x)$, called the *codistance*, between vertices of one and vertices of the other. This codistance is defined to satisfy the following condition:

if $\mathrm{codist}(x,y) = n$ *and if* y' *is adjacent to* y, *then* $\mathrm{codist}(x,y') = n \pm 1$;
furthermore if $n > 0$, *then* $+1$ *occurs for a unique such neighbour* y'.

Two vertices are called *opposite* if the codistance between them is 0. As in the case of generalised polygons, there is a canonical bijection between the neighbours of two opposite vertices, defined in this case by being at codistance 2, so opposite vertices have the same valency. In fact if the trees are thick (at least three edges per vertex), then the valency of opposite vertices being equal implies the trees are semi-homogeneous and isomorphic to one another—see the exercise below. In the example of the twin tree for the group $SL_2(k[t,t^{-1}])$, which appears as the first in a family of affine twin buildings discussed in Section 4, the valency of each vertex is $1+q$, where q is the cardinality of k.

Exercise 2.   If each vertex lies on at least three edges, prove that two vertices at even distance, or even codistance, from one another have the same valency. Conclude that in this case the two trees are isomorphic to one another.

Exercise 3.   Show that the definition of a twin tree given in terms of a codistance between vertices is equivalent to the definition given in Section 1 when W is the infinite dihedral group.

Twin Ends in Twin Trees.   In a tree having no vertices of valency 1, an *apartment* is a path infinite in both directions, and a *half-apartment* is a path infinite in one direction. Two half-apartments are said to have the same *end* if their intersection is a half-apartment; this is an equivalence relation on the set of half-apartments, and the equivalence classes are called the *ends* of the tree. Each apartment has two ends, and any two ends a and b span a unique apartment, denoted (ab); if x is a vertex and e an end, then x and e span a half apartment, denoted (xe). For a *twin tree* the twinning picks out a subset of ends that we define as follows.

   Take two vertices x and y in different components of the twin, and let $(y=y_0,y_1,y_2,\ldots)$ be a path along which the codistance from x increases monotonically. If $\text{codist}(x,y) > 0$, then this path is unique because there is only one direction in which codistance can increase, but in any case it is unique after the first step. The path determines an end of the tree, and the ends obtained in this way for each tree we call the *ends of the twinning*.

Exercise 4.   If the valency of each vertex is finite, show there are only countably many ends of the twinning.

Exercise 5.   Let γ be a path in one component of a twin tree, and x a vertex in the other component. Show that if there is no vertex of γ opposite x, then the codistance from x to the vertices of γ must increase or decrease monotonically along γ. Conclude that if γ is an apartment it contains a vertex opposite x.

(11.1) LEMMA.   *Given two pairs* (x,y) *and* (x',y') *of non-opposite vertices in a twin tree* (T_+,T_-), let e_+ *and* e_- (*respectively* e'_+ *and* e'_-) *denote the ends determined by* (x,y) (*respectively* (x',y')) *as above. If* $e'_+ = e_+$, *then* $e'_- = e_-$.

PROOF: Since $e'_+ = e_+$, the intersection $(xe_+)\cap(x'e_+)$ in T_+ is a path $(z_1,z_2, \ldots)$ leading towards e_+. Now suppose $e'_- \neq e_-$. Then there is an apartment A of T_- spanned by e'_- and e_-. If p and p' denote vertices of A that lie on (ye_-) and $(y'e_-)$ respectively, then the codistance from z_1 increases monotonically along (pe_-) and along $(p'e'_-)$. The apartment A is a union of paths (pe_-), $(p'e'_-)$, and (pp'), and along (pe_-) and $(p'e'_-)$ the codistance from z_2 is one greater than from z_1, raising the value of the codistance from z_2 to the end points of (pp'). By an obvious induction, for k sufficiently large there is no vertex of A opposite z_k on any of the paths (pe_-), $(p'e'_-)$, and (pp'), contradicting Exercise 5 and proving $e'_- = e_-$, as required.

(11.2) THEOREM.   *Given a twin tree* (T_+,T_-), *let* E_+ *and* E_- *denote the set of ends of the twinning for* T_+ *and* T_- *respectively. Then there is a canonical bijection between* E_+ *and* E_-.

PROOF: This follows from the lemma above.

Using the canonical bijection in (11.2) it is meaningful to talk of the *twin ends* of (T_+,T_-), meaning the ends of the twinning with the canonical identification between those for T_+ and those for T_-. Each pair of twin ends yields an apartment in each tree, and when these two apartments form a twin building in their own right we call their union a *twin apartment*.

Exercise 6.   Show that a pair of apartments, A_+ in T_+ and A_- in T_-, form a twin apartment if and only if each vertex of one is opposite a unique vertex of the other.

We shall now examine the concept of a twin apartment in any twin building.

3. Twin Apartments.

In a spherical building any two chambers lie in an apartment, and when they are opposite they lie in a *unique* apartment. The analogy in a twin building is that any two chambers lie in a *twin* apartment, and when they are opposite they lie in a unique twin apartment.

In order to define a twin apartment, we use the term *Coxeter building* to mean one in which each panel is on exactly two chambers (this is what we previously called a Coxeter complex). Given two Coxeter buildings Σ_+ and Σ_- of the same type W choose a pair of chambers (c_+,c_-) in (Σ_+,Σ_-); then for chambers x_+ in Σ_+, and y_- in Σ_-, define a codistance $w^*(x_+,y_-) = w(x_+,c_+).w(c_-,y_-)$. The original pair of chambers c_+ and c_- are opposite, and the twinning of Σ_+ and Σ_- is uniquely determined by this fact. This yields a *twin Coxeter building* of type W, and all such are isometrically equivalent.

Given a twin building $\Delta = (\Delta_+,\Delta_-)$, Tits [1992] defines a *twin apartment* to be any subset $A = (A_+,A_-)$ of chambers forming a twin Coxeter building. I shall call A_+ and A_- the *components* of A. In the spherical case each apartment in Δ_+ is naturally twinned with itself, but in the non-spherical case most apartments of Δ_+ fail to be part of a twin apartment—examples are easily seen using the twin trees of the last section.

To obtain a twin apartment, take opposite chambers x and y in Δ_+ and Δ_- respectively. Then define

$$A_+(x,y) = \{z \in \Delta_+ \,|\, w(x,z) = w^*(y,z)\},$$
$$A_-(x,y) = \{z \in \Delta_- \,|\, w(y,z) = w^*(x,z)\}.$$

Exercise 7.   Show that $A_+(x,y)$ and $A_-(x,y)$ are apartments in Δ_+ and Δ_- respectively.

The union $A(x,y) = A_+(x,y) \cup A_-(x,y)$ forms a twin apartment, as the following result shows.

(11.3) PROPOSITION.   *For a pair of apartments A_+ in Δ_+, and A_- in Δ_-, the following statements are equivalent.*

(i) (A_+,A_-) *is a twin apartment.*

(ii) $A_+ = A_+(x,y)$ *and* $A_- = A_-(x,y)$ *for some pair of opposite chambers x and y.*

(iii) *Each chamber of A_- (resp. A_+) is opposite exactly one chamber of A_+ (resp. A_-).*

PROOF: (i) $\Rightarrow$ (ii)   Let x and y be opposite chambers in (A_+,A_-), and let z in A_+ be in the same component as x. To prove (ii) we must show that $w(x,z) = w^*(y,z)$, so let $(x=x_0, \ldots,x_n=z)$ be a minimal gallery, of type $(s_1, \ldots,s_n)$, from x to z, and note that $w(x,x_k) = s_1.\ldots.s_k$ for $1 \le k \le n$. We know that $w^*(y,x_0) = 1$, and by induction it suffices to assume that $w^*(y,x_k) = s_1 \ldots s_k$, and prove that $w^*(y,x_{k+1}) = s_1 \ldots s_{k+1}$. Since (A_+,A_-) is a twin Coxeter building, this follows from (TB3) applied to y and the panel common to x_k and x_{k+1}.

(ii) ⇒ (iii) By definition, x and y are opposite one another and not opposite any other chamber of $A(x,y)$. If x' and y' are the unique chambers of $A(x,y)$ distinct from, and s-adjacent to, x and y respectively, then x' is opposite y', and it suffices to prove, by an obvious induction, that $A(x',y') = A(x,y)$. This will be true if one is contained in the other, so let z be any chamber of $A(x,y)$, and set $w = w(x,z) = w^*(y,z)$. We have $w(x',z) = sw$, because x, x' and z lie in a common apartment, and since y' is s-adjacent to y, we have $w^*(y',z) = w$ or sw; we must show it is sw. If w is s-extended on the left, or in other words if $|sw| < |w|$, this follows from (TB3) applied to z and the panel common to y and y', so suppose w is s-reduced on the left. Let z' be a chamber of $A(x,y)$ adjacent to z on a minimal gallery from x to z. So $w^*(y,z') = w(x,z') = s_1 \ldots s_{n-1}$, where $w = s_1 \ldots s_n$ in reduced form. By an obvious induction along a gallery from x to z, we may assume $w^*(y',z') = ss_1 \ldots s_{n-1}$. This has the same length as w, so if $w^*(y',z) = w$ then $w = ss_1 \ldots s_{n-1}$ by (TB2), and this reduced expression contradicts our assumption that w is s-reduced on the left.

(iii) ⇒ (i) Given (iii) we must verify axioms (TB1) to (TB3). Since (A_+,A_-) is a twin building, the first two axioms are automatically satisfied, so we need only check (TB3). Let y and z be s-adjacent chambers, and x a chamber in the other component of (A_+,A_-). Writing $w = w^*(x,y)$ we have $w^*(x,z) = w$ or ws, and must show that $w^*(x,z) = ws$. Suppose by way of contradiction that $w^*(x,z) = w$, and let $s_1 \ldots s_n$ be a reduced expression for w. Take a gallery of type $(s_n, \ldots, s_1)$ starting at y, and another gallery of the same type starting at z. These two galleries have different end chambers, both opposite x, which contradicts (iii) and completes the proof.

Exercise 8.   Any two chambers of a twin building lie in a twin apartment. [Hint: If x and y are in the same component, let $(s_1, \ldots, s_n)$ be the type of a minimal gallery from x to y; pick x' in the other component with $w^*(x',y) = s_1 \ldots s_n$, show x' is opposite x, and $y \in A(x,x')$. If x and y are in different components, let $w^*(x,y) = s_1 \ldots s_n$ (reduced) and let x' be the first term of a gallery of type $(s_1, \ldots, s_n)$ ending in y; show x' is opposite x, and $y \in A(x,x')$.]

Exercise 9.   Let A be a twin apartment, and x any chamber. Show that x is opposite at least one chamber of A, and x lies in A if and only if it is opposite a *unique* chamber of A.

4. An Example: Affine Twin Buildings.

Each spherical building can be canonically twinned with itself, as we saw in Section 1, but for affine buildings the situation is quite different; some can be twinned and some cannot. In Chapter 9 Section 2 affine buildings of type $\tilde{A}_{n-1}$ were obtained from a vector space of dimension n over a field K with a discrete valuation v onto the integers; as in that section, let k denote the residue field of K, in other words the field obtained from the valuation ring $O = \{a \in K \mid v(a) \geq 0\}$ by factoring out its unique maximal ideal $\{a \in K \mid v(a) \geq 1\}$. For example, when K is Q_p (the p-adic numbers) its valuation ring is $\mathbf{Z}_p$ (the p-adic integers), and k is the finite field of p elements. The vertices of the affine building are obtained as equivalence classes of O-lattices that span V, two lattices being equivalent if one is a scalar multiple of the other. In Chapter 9 we called them v-lattices.

For fields K having a discrete valuation there are two cases: either K and its residue field k have the same characteristic or, like the p-adics, K has characteristic 0 and k has characteristic p. Only in the equal characteristic case can the affine building have a twin, assuming it is not simply a product of trees. For example the irreducible p-adic affine buildings of rank at least 3 are not twinnable.

The Building at Infinity. In a non-spherical twin building $(\varDelta_+, \varDelta_-)$ "most" apartments A_+ of $\varDelta_+$ do not arise from a twin apartment (A_+, A_-). Those that do form an interesting structure; in the affine case, as A_+ ranges over the set of all such "twinnable" apartments for $\varDelta_+$, their union at infinity forms a spherical building, though this is not obvious because two chambers in this union do not necessarily lie in a common apartment $(A_+)^\infty$, where A_+ is twinnable. These two spherical buildings, one for $\varDelta_+$ and one for $\varDelta_-$, are canonically isomorphic. A brief discussion of the $\tilde{A}_2$ case is given below, while proofs in all cases are given in Ronan [2003]. Note that the subset of "twinnable" apartments is not a "system of apartments" in the sense of Chapter 10; two sectors, each lying in a twinnable apartment, do not necessarily contain sub-sectors lying in a common twinnable apartment.

In the $\tilde{A}_2$ case each apartment can be viewed as a tiling of the Euclidean plane by equilateral triangles, and the spherical apartments at infinity are hexagons (apartments of type A_2). Consider a twin apartment (A_+, A_-): each vertex x_+ of A_+ is opposite a unique vertex of A_-, and each sector of A_+ having vertex x_+ can be associated to a sector of A_- having vertex x_- so that both are in the "same direction" (see [loc. cit.] for details). Containment of one sector inside another is reversed between A_+ and A_-; the larger the sector in A_+ the smaller the sector in A_-. This bijection between sectors of A_+ and A_- yields a canonical bijection between chambers of $(A_+)^\infty$ and $(A_-)^\infty$, and for each chamber at infinity one can show that its image under this bijection is independent of the twin apartment containing it. This yields a canonical bijection between the buildings at infinity obtained from the twinning.

Note that if we compactify the affine apartments A_+ and A_- by adjoining $(A_+)^\infty$ and $(A_-)^\infty$ respectively, and then identify $(A_+)^\infty$ with $(A_-)^\infty$, we obtain a space that is topologically a sphere. This allows us to glue together the two components of the twin building by gluing along the boundaries of the twin apartments. For example in the rank 2 affine case, where $\varDelta_+$ and $\varDelta_-$ are trees, each apartment can be thought of as the real line with integer points as vertices. A twin apartment is a union of two such lines with a canonical identification of their ends. Compactifying these lines by adjoining their ends, and treating their union topologically as a circle, gives a structure that has more in common with a generalised polygon than a tree.

The Affine Twin Building for $\mathrm{SL}_n(k[t,t^{-1}])$.   Let $K = k(t)$ denote the field of rational functions on the projective line over k. Take two rational points on this line, and let A denote the ring of functions in $k(t)$ having poles only at these two points. Without loss of generality the points are 0 and ∞, in which case $A = k[t,t^{-1}]$. The order of a zero or a pole at 0 (or at ∞) gives, in the usual way, a discrete valuation that we denote by v_+ (or v_-).

Given an n-dimensional vector space V over K, as above, the two valuations v_+ and v_- yield two affine buildings $\varDelta_+$ and $\varDelta_-$ of type $\tilde{A}_{n-1}$, as in Chapter 9 Section 2. These buildings can be twinned with one another, and each free A-module M that is spanned by a basis of V yields a twinning.

In Chapter 9 the affine building for a given valuation was described in terms of its vertices, but the concept of a twinning has been framed in terms of chambers. At present nothing in the literature expresses a twinning of two affine buildings in terms of a "codistance" between vertices (except in the twin tree case), but a twinning is uniquely determined by its pairs of opposite chambers, and these in turn are determined by the pairs of opposite vertices. We now provide a necessary and sufficient condition for two vertices to be opposite, given the A-module M mentioned above.

Let O_+ and O_- denote the local rings at 0 and ∞—they comprise the functions in K having no pole at 0 and ∞ respectively. One has

$$O_+ \cap A = k[t]$$
$$O_- \cap A = k[t^{-1}]$$

Each vertex x_+ of Δ_+ can be represented by an O_+-lattice L_+, and each vertex y_- of Δ_- by an O_--lattice L_-, these lattices being unique up to multiplication by an element of K. For the twinning, let $X_+ = L_+ \cap M$, and $Y_- = L_- \cap M$, so X_+ is a $k[t]$-module, and Y_- is a $k[t^{-1}]$-module. The intersection $X_+ \cap Y_-$ is a k-module of finite dimension, and if we replace X_+ by $t^n X_+$ this k-module increases or decreases in dimension according to whether n decreases or increases; for example if $z \in X_+ \cap Y_-$ then both z and $t^{-1}z$ lie in $t^{-1}X_+ \cap Y_-$, from which one can see that if $X_+ \cap Y_- \neq 0$, then replacing $n = 0$ by $n = -1$ increases the dimension. Replacing X_+ by $t^n X_+$ for some suitable n does not change the vertex x_+ of Δ_+, but it allows us to assume that $X_+ \cap Y_-$ contains a basis for V, while $tX_+ \cap Y_-$ does not. With this choice of X_+ and Y_- the vertices x_+ and y_- are opposite in the twinning determined by M if and only if $tX_+ \cap Y_- = 0$. Another way of putting this is that given a basis for M, the O_+ and O_- lattices spanned by this basis yield vertices that are opposite, and all pairs of opposite vertices are obtained in this way.

For twin trees, Ronan and Tits [1994] show that this opposition relation between vertices yields a twinning. A verification in higher rank is given by Abramenko and Van Maldeghem [2001] who study this example in detail and show directly how it leads to a twin BN-pair.

The group $G = \mathrm{SL}(M) \cong \mathrm{SL}_n(k[t,t^{-1}])$ acts transitively on pairs of opposite vertices, and on pairs of opposite chambers, which implies it operates transitively on the set of twin apartments. The building at infinity described above, obtained using the twin apartments of Δ_+ (or Δ_-) is isomorphic to the spherical building for $\mathrm{SL}_n(k(t))$, which is of course transitive on its set of apartments. The subgroup G preserves the subset of apartments $(A_+)^\infty$ arising from the twinnable apartments A_+ of (Δ_+, Δ_-).

5. Residues, Rigidity, and Proj.

As mentioned in Section 1, two residues in different components of a twin building are said to be opposite one another if they have the same type and contain chambers that are opposite. (In a single spherical building if one has type J its opposite has type $w_0 J w_0$, where w_0 is the longest word of W, but in the canonical twinning they have the same type). If R and S are opposite residues then (R,S) is a twin building in its own right, a fact that can be verified by checking axioms (TB1) to (TB3). The use of opposites is a powerful technique in the theory of twin buildings, as it is for spherical buildings, and one can use opposite rank 1 residues to prove the following rigidity theorem, which is an analogue of (6.4) for spherical buildings.

(11.4) THEOREM. *Let Δ be a thick twin building, and (b,c) a pair of opposite chambers in Δ. Then the only isometry of Δ fixing all chambers of $E_1(c) \cup \{b\}$ is the identity.*

PROOF: First note that if R and S are opposite residues of type $\{s\}$, then the property of being at codistance s sets up a canonical bijection between R and S, so an isometry fixing all chambers of R and one chamber of S must fix all chambers of S. Therefore an isometry fixing all chambers of $E_1(x) \cup \{y\}$ must fix all chambers of $E_1(x) \cup E_1(y)$. By induction along a gallery from c, or from b, it suffices to prove that if x' is s-adjacent to x, then an isometry fixing all chambers of $E_1(x) \cup E_1(y)$ must fix all chambers of $E_1(x') \cup E_1(y')$, for some y' opposite x'. If R denotes the s-residue containing y, then one chamber of R is at codistance s from x, and one at codistance s from x', so by thickness there is a chamber y' of R different from these, and therefore opposite both x and x'. Since our isometry fixes $E_1(x) \cup \{y'\}$, it fixes all chambers of $E_1(x) \cup E_1(y')$; in particular it fixes $\{x'\} \cup E_1(y')$, hence all chambers of $E_1(x') \cup E_1(y')$, completing the inductive step.

Remark. The canonical bijection between opposite residues R and S of rank 1 is a feature of opposite residues of higher rank, provided they have spherical type. Indeed if R and S are spherical of type J, then W_J is a finite Coxeter group, and it has a longest word r_J. Using (TB3) and (TB2) it is straightforward to show that each chamber of R is at codistance r_J from a unique chamber of S, and vice versa. This gives a canonical bijection between R and S.

Proj. More generally, given a chamber x and a spherical residue S in different components of a twin building, there is a unique chamber of S at maximal codistance from x; we denote it $\mathrm{proj}_S x$. This is analogous to the case of x and S being in the same component, where $\mathrm{proj}_S x$ is the unique chamber of S at minimal distance from x. Before defining proj_S in terms of maximal codistance, we refer to an element w of W as *J-reduced on the right* if it is r_j-reduced on the right for all j in J, or in other words if $|wr_j| > |w|$ for all j in J. A similar definition applies with "right" replaced by "left".

Exercise 10.   For any w in W, and any type J, show that $w = u_J v_J$ where u_J is J-reduced on the right and $v_J \in W_J$. Both u_J and v_J are uniquely determined by w. [Hint: treat 1 and w as chambers of the Coxeter complex, let S denote the J-residue containing w, and consider $\mathrm{proj}_S 1$].

Given a spherical residue S of type J in a twin building Δ, its Coxeter group W_J has a unique longest word r_J, and for a chamber x in the other component of Δ, we use r_J in defining $\mathrm{proj}_S x$. When S is *not* spherical, W_J has no longest word and $\mathrm{proj}_S x$ does not exist.

(11.5) PROPOSITION.   *Let x be a chamber in one component of a twin building, and S a spherical residue of type J in the other component. Then as y ranges over the chambers of S:*

(i) *the length of $w^*(x,y)$ takes its maximum value at a unique chamber $p = \mathrm{proj}_S x$ in S, and $w^*(x,p) = u r_J$ where u is J-reduced on the right, and r_J is the longest word in W_J;*

(ii) $w^*(x,y) = w^*(x,p)w(p,y)$ *and* $|w^*(x,y)| = |w^*(x,p)| - |w(p,y)|$;

PROOF: Treat W as a Coxeter building: as y ranges over S, $w^*(x,y)$ ranges over a J-residue V in W, and we let $u = \mathrm{proj}_V 1$. Then u is J-reduced on the right, and $w^*(x,y) = uv$, where v is in W_J. Let p be a chamber in S with $w^*(x,p) = u r_J$, and for any chamber y in S let $(s_1, \ldots, s_n)$ be the type of a minimal gallery from p to y, so $w(p,y) = s_1 \ldots s_n$. Then $|r_J s_1 \ldots s_k| = |r_J| - k$, so $|u r_J s_1 \ldots s_k| = |u r_J| - k$, and by (TB3) and an obvious induction, $w^*(x,y) = u r_J s_1 \ldots s_n = w^*(x,p)w(p,y)$. This verifies (ii), and shows that $y = p$ is the unique chamber for which $w^*(x,y)$ takes its maximum value.

Exercise 11.   Let x and S be as in (11.5). If A is any twin apartment containing both x and a chamber of S, show that $\mathrm{proj}_S x$ lies in A. Moreover if x' is the chamber of A opposite x, then $\mathrm{proj}_S x'$ is opposite $\mathrm{proj}_S x$ in S.

Exercise 12.   Let R and S be opposite residues of spherical type J. Show that $\mathrm{proj}_R|S$ and $\mathrm{proj}_S|R$ are inverse bijections sending r-adjacent chambers to s-adjacent chambers, where $rr_J = r_J s$ (r_J being, as above, the longest word of W_J).

6. 2-Spherical Twin Buildings.

The rank 2 residues of a building are either generalised polygons or trees. When they are generalised polygons the building is called *2-spherical*.

(11.6) THEOREM.   *In a 2-spherical twin building whose diagram has no direct factors of rank 2, each spherical residue of rank 2 is either a generalised 2-gon or a Moufang polygon.*

PROOF: Tits [1992] outlined the ingredients for proving this, and a complete proof is given in Ronan [2000; Theorem 4].

(11.7) COROLLARY.   *In a 2-spherical twin building whose diagram has no direct factors of rank 2, each rank 2 residue is either a tree or generalised m-gon for m = 2, 3, 4, 6, or 8.*

PROOF: This follows from the previous theorem along with the results of Tits [1976/79] and Weiss [1979].

Extending Isometries—Local-to-Global Structure.   Recall from Chapter 6 that $E_2(c)$ means the union of the rank 2 residues containing the chamber c; we shall refer to it as the *local structure* of the building concerned. Tits [1974] showed that the local structure of a spherical building uniquely determines its global structure, a fact expressed earlier as Theorem (6.6). This result admits a generalization to 2-spherical twin buildings $\Delta = (\Delta_+, \Delta_-)$, and in (11.9) we show that the local structure determines the global structure of Δ_+ (and Δ_+). For the global structure of Δ itself, see (11.11). First we need the following lemma.

(11.8) LEMMA.   *Given two pairs of opposite chambers (x,z) and (x',z') in twin buildings that are 2-spherical, any isometry φ from $E_2(x) \cup \{z\}$ to $E_2(x') \cup \{z'\}$ extends uniquely to an isometry from $E_2(x) \cup E_2(z)$ to $E_2(x') \cup E_2(z')$.*

PROOF: The uniqueness is straightforward to see, using the remark in Section 5: if S is a rank 2 residue of type J containing z, then each chamber of S is $\text{proj}_S y$ for a unique chamber y lying in the J-residue containing x, so the extension must send $\text{proj}_S y$ to $\text{proj}_{S'}(\varphi y)$ where S' is the residue of type J containing z'. The fact that it preserves codistances between $E_2(x)$ and $E_2(z)$ is proved in (6.2) of Ronan [2000].

Remark.   Given two pairs of opposite chambers (x_1, z_1) and (x_2, z_2) where x_2 is s-adjacent to x_1, and z_2 is s-adjacent to z_1, Lemma (11.8) implies that an isometry φ from $E_2(x_1) \cup E_2(z_1)$ to $E_2(x'_1) \cup E_2(z'_1)$, yields a unique isometry from $E_2(x_2) \cup E_2(z_2)$ to $E_2(x'_2) \cup E_2(z'_2)$, where $(x'_2, z'_2) = (\varphi x_2, \varphi z_2)$. We use this in proving the following theorem, generalising (6.6), which can be equally well phrased in terms of a pair (c,d) of opposite chambers, rather than the equivalent notion of a chamber c and a twin apartment A containing it, as it was in Chapter 6.

(11.9) THEOREM. *Let (c,d) and (c',d') be pairs of opposite chambers in 2-spherical twin buildings (Δ_+,Δ_-) and (Δ'_+,Δ'_-) respectively. Then given any isometry φ_c from $E_2(c)\cup\{d\}$ onto $E_2(c')\cup\{d'\}$ there is an isometry from Δ_+ onto Δ'_+ that restricts to φ_c on $E_2(c)$. In particular the local structure determines the global structure of Δ_+.*

PROOF: Let A_- be an apartment of Δ_- containing d, and for any chamber x in Δ_+ let πx denote the unique chamber of A_- defined by $w(d,\pi x) = w^*(d,x)$. Then x is opposite πx, and $\pi x = d$ whenever x is opposite d. By (11.8) when x is adjacent to c the isometry φ_c with domain $E_2(c)\cup\{d\}$ yields an isometry φ_x with domain $E_2(x)\cup\{\pi x\}$, and an obvious induction along galleries from c gives an isometry φ_x for each chamber x in Δ_+. This isometry is independent of the gallery because a circuit in a rank 2 residue does not change the isometry (see Ronan [2000; (6.7)] for a proof of this fact). We therefore obtain a well-defined isometry with domain Δ_+, as required.

Remark. The apartment A_- can be chosen freely—it does not need to be part of a twin apartment—but given A_-, the extension is unique. Using a different apartment it is not clear that one obtains the same extension, but this uncertainty is avoided if the set of chambers opposite d is connected, and we now introduce the following notation.

Notation. Given a chamber c in a twin building, c^{op} will mean the set of chambers opposite c.

Condition (co). A twin building is said to satisfy (co) if for any chamber c the set of chambers opposite c is connected, meaning that between any two chambers of c^{op} there is a gallery lying entirely in c^{op}.

Exercise 13. Show that the two components of a 2-spherical twin building are isomorphic.

Exercise 14. Prove that if each rank 2 residue satisfies (co) then so does Δ. [Hint: given a gallery $\gamma = (x_0,\ldots,x_n)$ between two chambers opposite c, set $w_i = w^*(c,x_i)$, and if one of these is not the identity take $|w_{i-1}| < |w_i|$ where $|w_i|$

is maximal, and modify γ in the rank 2 residue containing x_{i-1}, x_i and x_{i+1} so as to eliminate x_i].

Exercise 15.   With the notation of (11.9) show that an isometry from $\varDelta_+$ onto $\varDelta'_+$ sending d^{op} to $(d')^{\mathrm{op}}$ is an isometry from $\varDelta_+\cup\{d\}$ onto $\varDelta'_+\cup\{d'\}$, and that an isometry from $\varDelta_+\cup\{d\}$ onto $\varDelta'_+\cup\{d'\}$ is uniquely determined by its action on d^{op}.

(11.10) THEOREM.   *Let (c,d) and (c',d') be pairs of opposite chambers in 2-spherical twin buildings $(\varDelta_+,\varDelta_-)$ and $(\varDelta'_+,\varDelta'_-)$ that satisfy* (co). *Then any isometry from $E_2(c)\cup\{d\}$ onto $E_2(c')\cup\{d'\}$ extends to a unique isometry from $\varDelta_+\cup\{d\}$ onto $\varDelta'_+\cup\{d'\}$.*

PROOF: By (11.8) an isometry from $E_2(c)\cup\{d\}$ onto $E_2(c')\cup\{d'\}$ extends uniquely to an isometry φ_c from $E_2(c)\cup E_2(d)$ to $E_2(c')\cup E_2(d')$, and for any chamber x adjacent to c and opposite d it induces a unique isometry φ_x from $E_2(x)\cup E_2(d)$ to $E_2(\varphi_c x)\cup E_2(d')$. This implies, by condition (co) and an obvious induction along galleries in d^{op}, that any extension to an isometry from $\varDelta_+\cup\{d\}$ to $\varDelta'_+\cup\{d'\}$ is uniquely determined on d^{op}, and hence by the exercise above is unique. In order to prove there is such an extension, let φ_+ denote the isometry from $\varDelta_+$ to $\varDelta'_+$ as in (11.9). If x is opposite d then by the exercise above it suffices to prove that $\varphi_+(x)$ is opposite d'. But in the proof of (11.9), φ_+ is a combination of isometries φ_x with domains $E_2(x)\cup\{\pi x\}$, each obtained by induction along a gallery from c to x, and φ_x is independent of the gallery. When x is opposite d, then $\pi x = d$ and by (co) there is a gallery from c to x lying entirely within d^{op}; along such a gallery, each local isomorphism sends d to d', so $\varphi_x(d) = d'$. This implies $\varphi_+(x)$ is opposite d', as required.

Remarks and Definition.   Condition (co) is almost always satisfied in a 2-spherical twin building. This is because a twin building satisfies (co) if each of its rank 2 residues does (see Exercise 14), and by (11.6) these rank 2 residues are either generalised 2-gons or Moufang polygons. Each generalised 2-gon satisfies (co), and Abramenko and Van Maldeghem [1999] have shown the same is true for all Moufang polygons except those belonging to four finite groups of Lie type, namely $\mathrm{Sp}_4(2)$, $\mathrm{G}_2(2)$, $\mathrm{G}_2(3)$ and ${}^2\mathrm{F}_4(2)$. We shall call a 2-spherical twin building *non-fragile* if it contains none of these rank 2 residues.

The following theorem gives a local-to-global theorem for non-fragile, 2-spherical twin buildings.

(11.11) THEOREM. *Let (c,d) and (c',d') be pairs of opposite chambers in 2-spherical twin buildings Δ and Δ' respectively, and assume Δ is non-fragile. Then any isometry from $E_2(c) \cup \{d\}$ onto $E_2(c') \cup \{d'\}$ extends uniquely to an isometry from Δ to Δ'. In particular the local structure of Δ determines its global structure.*

PROOF. This theorem is proved in Mühlherr and Ronan [1995].

In the fragile case there are quite likely non-isometric twinnings, but in the non-fragile case this theorem can be used in the classification of 2-spherical twin buildings by classifying the local data. This difficult problem has been dealt with by Mühlherr [1999] and [2002].

7. The Moufang Property and Root Group Data.

The concept of a root in a single apartment extends to twin apartments in the following way. Take a twin apartment $A = (A_+, A_-)$ and a half-apartment a in A_+. The chambers of A_- opposite those of a form a half apartment a^{op} whose wall ∂a^{op} is opposite the wall ∂a of A_+. Let α denote the union of a and (the complement of a^{op} in A); it is called a *root* (or *twin root*) of the twin apartment (A_+, A_-). Like a root of a spherical apartment, it contains no pair of opposite chambers, and every chamber of A is opposite a chamber of α. Let α_+ denote $\alpha \cap A_+$ and α_- denote $\alpha \cap A_-$.

The group of automorphisms fixing α and all chambers on an interior panel of α (i.e. a panel not on the boundary $\partial \alpha$) is called a *root group* and denoted U_α. As with roots in spherical buildings, U_α acts freely on the set of twin apartments containing α; this is a simple consequence of (11.4). If U_α is transitive on this set, we call it a *full* root group, and if all root groups are full the twin building is called *Moufang*. As with spherical buildings, in Chapter 6 Section 2, it suffices to assume that U_α is a full root group for all roots α in a given twin apartment A, in which case the group generated by these U_α is transitive on the set of twin apartments.

Before proving our next theorem, showing that non-fragile, 2-spherical buildings are Moufang, here are some definitions and exercises.

Definition. Given a root α, a panel of α is called *interior* if it does not lie on the boundary $\partial \alpha$, and a chamber is *interior* to α if all its panels are; chambers or panels of α that are not interior are called *boundary* chambers or panels. A residue is called *interior* to α if it has no panels on the boundary, and a chamber lies in the *second interior* of α if all rank two residues containing it are interior.

In the following three exercises α is a half-apartment in a single Coxeter building Σ.

Exercise 16. If a rank 2 residue R of Σ is interior to α and contains two non-adjacent boundary chambers x and y, then all chambers of R are boundary chambers, and their boundary panels all have the same type. [Hint: Let J be the type of R, let s and t be the types of boundary panels for x and y, and let x' and y' respectively be the other chambers on these panels. If w is the type of a minimal gallery from x to y, then swt is the type of a gallery from x' to y' crossing $\partial \alpha$ twice, and the exchange property—Exercise 4 in Chapter 2—shows that $s = t$; now consider the $\{s, J\}$ residue containing R].

Exercise 17. Assume the diagram has no isolated nodes. Show that given two interior chambers in α, there is a gallery from one to the other lying in the interior of α. [Hint: use induction on the distance between two interior chambers].

Exercise 18. Assuming each connected component of the diagram has rank at least 3, show that α contains a chamber in its second interior.

Exercise 19. Let α be a root of a twin Coxeter building Σ. If R_+ is a boundary residue of α_+, and R_- is the opposite residue of Σ, show that R_- is a boundary residue of α_-, and proj induces a bijection between $R_+ \cup \alpha_+$ and $R_- \cup \alpha_-$.

Exercise 20. Let Δ be a 2-spherical twin building for which each connected component of the diagram has rank at least 3. Let α be a twin root, and x and y adjacent interior chambers of α. If an automorphism of Δ fixes α and $E_1(x)$, then it fixes $E_1(y)$. [Hint: consider the rank 2 residues containing x and y, and use (6.5) and (11.6)].

(11.12) THEOREM. *If Δ is a non-fragile, 2-spherical twin building for which each connected component of the diagram has rank at least 3, then Δ is Moufang.*

PROOF: Let α be a (twin) root, and pick a chamber c in the second interior of α, so if A is a twin apartment containing α then $E_2(c) \cap A \subset \alpha$. Let A and A' be two such apartments and let d and d' be their chambers opposite c, so the identity map on $E_2(c)$ extends to an isometry from $E_2(c) \cup \{d\}$ to $E_2(c) \cup \{d'\}$.

By the local-to-global theorem (11.11) this extends to an isometry g from Δ to itself, and since g sends $\{c,d\}$ to $\{c,d'\}$, it sends A to A' and hence fixes α. It remains to show that g fixes all chambers on panels interior to α.

Let α_+ and α_- denote the two components of α, with c in α_+, and let Δ_+ and Δ_- denote the corresponding components of Δ. If x is an interior chamber of α_+, then by Exercise 17 there is a gallery from c to x lying in the interior of α, and by induction along this gallery Exercise 20 implies that g fixes $E_1(x)$. Every interior panel lies in a rank 2 interior residue R; this contains an interior chamber x, and since g fixes $\alpha \cap R$ and $E_1(x)$, it fixes all chambers of R. This deals with interior panels of α_+ and we now turn to α_-.

Let R_- be a rank 2 residue of Δ_- containing a panel of $\partial\alpha$, and let R_+ denote the residue of the same type containing the opposite panel of $\partial\alpha$. Then R_- and R_+ are opposite residues and by Exercise 12 proj induces a canonical bijection between them. If π is an interior panel in R_-, then $\text{proj}_{R_+} \pi$ is an interior panel in R_+, and since g fixes all chambers on $\text{proj}_{R_+} \pi$ it fixes all chambers on π. In particular if b is a boundary chamber of α_-, whose boundary panel has type t, then g fixes the chambers on each panel of b having type $\neq t$. Now let a be a chamber of α_- that is s-adjacent to b, with $m_{st} \geq 3$. Then a is an interior chamber of α_-, and we claim that g fixes $E_1(a)$. If R_- denotes the $\{s,t\}$-residue containing a, then R_- contains a panel of $\partial\alpha$, namely the t-panel of b, so by the argument above g fixes all chambers on the interior panels of R_-, and in particular the s- and t-panels of a. If $r \neq s,t$ then the $\{r,s\}$-residue S containing a is an interior residue of α in which g fixes the apartment $S \cap \alpha$, along with the r- and s-panels of b; therefore g is the identity on S, and in particular fixes all chambers on the r-panel of a. This shows that g fixes $E_1(a)$, and hence by Exercise 20, $E_1(x)$ for every interior chamber of α. An interior panel not on an interior chamber lies in a boundary residue of rank 2, and by the argument at the start of this paragraph g fixes all chambers on such panels. Thus g fixes all interior panels of α_-, as required.

Root Group Data.   Tits [1992] gives a set of axioms for what he calls an *RGD-system*, where RGD stands for "root group data". The axioms, given below, are similar to those in Chapter 6 for a Moufang building.

Given a Coxeter group W with a distinguished set of generators S, we regard it as a chamber system as in Chapter 2. Let Φ be its set of roots (or half-apartments), Φ_+ the set of "positive" roots, meaning those containing the identity element, and Φ_- the complement of Φ_+ in Φ. If α is a root then $-\alpha$

denotes the complementary set of chambers; one lies in Φ_+ and the other in Φ_-. If s is a distinguished generator for W, let α_s denote the *fundamental* root defined by $\{w \| sw\| = |w|+1\}$—it lies in Φ_+.

Tits [loc. cit.] considers systems comprising a group G along with subgroups U_α indexed by the roots α of W satisfying the axioms below. The term *prenilpotent* for a pair of roots $\{\alpha,\beta\}$ is given in Chapter 6 Section 4, as is the set (α,β) for such a pair. In these axioms H denotes the intersection of the normalisers of the U_α in G, and U_+ denotes the subgroup of G generated by the U_α for $\alpha \in \Phi_+$.

(RGD 0) $U_\alpha \neq \{1\}$ for all α in Φ;

(RGD 1) If $\{\alpha,\beta\}$ is a prenilpotent pair of distinct roots the commutator $[U_\alpha, U_\beta] \leq U_{(\alpha,\beta)}$;

(RGD 2) For each distinguished generator s, and u in $U_{\alpha_s} - \{1\}$ there exist elements u' and u'' in $U_{-\alpha_s} - \{1\}$ such that the product $m(u) = u'uu''$ conjugates U_β onto $U_{s(\beta)}$ for all β in Φ;

(RGD 3) For each distinguished generator s, $U_{-\alpha_s} \not\subset U_+$;

(RGD 4) The group G is generated by H and the U_α's.

The elements $m(u) = u'uu''$ were described earlier in Chapter 6, and u determines both u' and uu''. As above $U_+ = \langle U_\alpha \mid \alpha \in \Phi_+ \rangle$, and one defines $U_- = \langle U_\alpha \mid \alpha \in \Phi_- \rangle$. One also defines $B_+ = HU_+$, $B_- = HU_-$, and $N = \langle H, m(u) \mid u \in U_{\alpha_s}, s \in S \rangle$. There is a unique homomorphism from N to W taking H to 1 and $m(u)$ to s for u in U_{α_s}, and the pairs (B_+, N) and (B_-, N) are twin BN-pairs for G. Thus an RGD-system yields a twin building, though a proof of this fact is non-trivial. However, Abramenko and Brown [2008] give a careful analysis of RGD-systems, and prove the existence of a twin BN-pair.

Root group data can be defined for all types of buildings, and in particular for those that are not 2-spherical. In this case some residues will yield twin trees, and B. Rémy and the author [2006] construct root group data for twin buildings in which all irreducible rank 2 residues are twin trees; the vertex stabiliser in one building yields a lattice acting on the other building and when the trees arise from fields of different characteristics these lattices exhibit unusual properties.

For a tree there are numerous possibilities for root data, even when it has a given finite valency, and to end this chapter, we now return to twin trees.

8. Twin Trees Again.

We have already seen that in a thick twin tree (T_+, T_-) each of T_+ and T_- must be semi-homogeneous. A natural question is whether every semi-homogeneous tree admits a twinning. Comparing twin trees to generalised polygons suggests this may not be the case, but in fact twin trees are abundant for any pair of valences, and the following theorem is proved by Ronan and Tits [1999].

(11.13) THEOREM. *A thick semi-homogeneous tree whose set of vertices has cardinality α admits 2^α isomorphism classes of twinnings, and among these 2^α have trivial automorphism group.*

This theorem says that rigid twin trees are at least as abundant as any other type, and we turn now to the opposite case, that of Moufang twin trees.

Moufang Twin Trees. A (twin) root in a twin tree can be thought of as half a twin apartment containing one of its two twin ends. More precisely, if e and f are ends spanning a twin apartment A, then each pair of opposite vertices x and y in A determines two roots: $\alpha = (xe) \cup (ey)$ containing e, and $-\alpha = (xf) \cup (fy)$ containing f. The *boundary* of α, denoted $\partial\alpha$, is $\{x, y\}$.

As before, the group of automorphisms fixing α and every edge containing a vertex of α-$\partial\alpha$ is denoted U_α, and we have already observed that it acts freely on the set of twin apartments containing α. When this action is transitive we call it a full root group, and if U_α is a full root group for all roots α we call the twin tree *Moufang*.

The set of roots in a twin apartment splits naturally into two subsets, one for each end. Roots involving the same end can be naturally indexed by the integers: $\ldots, \alpha_{n-1}, \alpha_n, \alpha_{n+1}, \ldots$ where the boundary vertices of α_n are adjacent to those of α_{n+1} for each n. Given α_m and α_n, the roots α_i as i ranges between m and n are precisely those roots containing $\alpha_m \cap \alpha_n$. As in the case of spherical buildings, in Chapter 6 Section 3, there are commutator relations between root groups, and in the following theorem U_n denotes the root group for the root α_n.

(11.14) THEOREM. *Given $m < n$, the group $[U_m, U_n]$ generated by the commutators $ghg^{-1}h^{-1}$, for g in U_m and h in U_n, is contained in the product $U_{m+1} \cdots U_{n-1}$, this product being defined as $\{1\}$ if $n = m + 1$.*

PROOF: Exercise.

The twin tree for $SL_2(k[t,t^{-1}])$ in Section 3 is Moufang, its root groups being isomorphic to the additive group of k. Moufang twin trees deserve a more detailed study, though they may not be susceptible to a classification in the same way as Moufang polygons.

APPENDIX 1
Moufang Polygons

This appendix has three sections. The first deals with the function $u \to m(u)$ introduced in Chapter 6, and proves the first statement of Lemma (7.3). The second section deals with Moufang planes, and derives the formula for the natural blueprint, used in Chapter 8. The third section proves the theorem (6.9) due to J. Tits and R. Weiss, that for a Moufang (generalised) d-gon, $d = 3$, 4, 6 or 8.

1. The m-function.

We recall from Chapter 6 that for any Moufang polygon (or indeed any Moufang building, given a root α in the apartment Σ, and given any $u \in U_\alpha - \{1\}$ there are unique elements $v, v' \in U_{-\alpha}$ such that

$$m(u) = vuv' \in N.$$

Abusing notation slightly, we let m denote the function sending $u \in U_\alpha - \{1\}$ to $m(u) \in N$, and let v, v' denote the functions interchanging U_α with $U_{-\alpha}$, where $v(u) = v$ and $v'(u) = v'$ above.

If c, c' denote respectively chambers of $\alpha, -\alpha$ which are adjacent (i.e. share a panel of $\partial\alpha$), then v is uniquely determined by sending $u(c')$ to c, and v' by sending c to $u^{-1}(c)$ - see Figure 1; remember that group action is on the left.

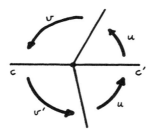

Figure 1

(A.1) LEMMA. *(i)* $m(u) = m(v) = m(v')$, *where* $v = v(u)$, $v' = v'(u)$,
(ii) $v(v'(u)) = v'(v(u)) = u$.

PROOF: Since vu sends c' to c we have

$$(vu)U_{-\alpha}(vu)^{-1} = U_\alpha$$

and hence

$$x = vuv'(vu)^{-1} \in U_\alpha.$$

Thus $xvu = vuv' \in N$, and therefore $m(v) = xvu = m(u)$, and $v'(v(u)) = u$.

Similarly:

$$y = (uv')^{-1}v(uv') \in U_\alpha.$$

Therefore $uv'y = vuv' \in N$, hence $m(v') = m(u)$, and $v(v'(u)) = u$.   □

Notation. To avoid cumbersome notation we shall write $^g x$ to mean gxg^{-1}, and (occasionally) x^g to mean $g^{-1}xg$.

(A.2) LEMMA. $m(u^{-1}) = m(u)^{-1}$ and $m(^n u) = {}^n m(u)$ for $n \in N$.

PROOF: Both these equations are immediate consequences of the fact that $U_{-\alpha}uU_{-\alpha} \cap N$ is a single element, namely $m(u)$; for instance both $m(u^{-1})$ and $m(u)^{-1}$ lie in $U_{-\alpha}u^{-1}U_{-\alpha} \cap N$.   □

Now consider a generalised d-gon (d for diameter). Let U_r, $r(\mod 2d)$, be the root groups in a natural cyclic order for the roots in a fixed apartment; in particular $U_{-r} = U_{r+d}$. As before, the commutator $[x, y] = xyx^{-1}y^{-1}$.

Given $e_1 \in U_1 - \{1\}$ and $e_d \in U_d - \{1\}$ we know by (6.12) that we may write

$$[e_1^{-1}, e_d] = e_2 \ldots e_{d-1}$$

where $e_r \in U_r$. Now define

$$e_{r+d} = v'(e_r) \in U_{r+d}$$

so the e_r are defined for all r. We also set

$$n_r = m(e_r).$$

(A.3) LEMMA. $e_{r+1} = n_r^{-1} e_{d+r-1} n_r$.

PROOF: Set $v = v(e_1) \in U_{d+1}$.

Then

$$
\begin{aligned}
e_{d+1} n_1^{-1} v &= e_{d+1} v'(e_1)^{-1} e_1^{-1} \\
&= v'(e_1) v'(e_1)^{-1} e_1^{-1} \qquad \text{(definition of } e_{d+1}) \\
&= e_1^{-1}. \qquad\qquad\qquad (*)
\end{aligned}
$$

Therefore

$$
\begin{aligned}
e_2 \ldots e_{d-1} e_d &= [e_1^{-1}, e_d] e_d && \text{(by definition)} \\
&= [e_{d+1} n_1^{-1} v, e_d] e_d, && \text{by } (*) \\
&= [e_{d+1} n_1^{-1}, e_d] e_d, && \text{since } [U_{d+1}, U_d] = 1 \\
&= e_{d+1} n_1^{-1} e_d n_1 e_{d+1}^{-1} \\
&= e_{d+1} x e_{d+1}^{-1}, && \text{where } x = n_1^{-1} e_d n_1 \in U_2 \\
&= x[x^{-1}, e_{d+1}].
\end{aligned}
$$

Since $x \in U_2$, $[x^{-1}, e_{d+1}] \in U_{[3,d]}$ by (6.12). Therefore by the uniqueness of the decomposition of the product $U_2 \ldots U_d$, we have

$$e_2 = x = n_1^{-1} e_d n_1$$

which is the $r = 1$ case of the lemma. Moreover this shows that

$$e_3 \ldots e_d = [e_2^{-1}, e_{d+1}].$$

Therefore we can proceed as above with all indices increased by 1, obtaining $e_3 = n_2^{-1} e_{d+1} n_2$, and $[e_3^{-1}, e_{d+2}] = e_4 \ldots e_{d+1}$, and proceed inductively, completing the proof. $\qquad\qquad\square$

(A.4) LEMMA. *(i)* $n_{r+d} = n_r$.
(ii) $n_r n_{r+1} = n_{r-1} n_r$.

PROOF: (i) Using (A.1)(i) for the middle equality, one has

$$n_{r+d} = m(v'(e_r)) = m(e_r) = n_r.$$

(ii) Using (A.2) and (A.3) for the first equality, and (i) for the second, one has

$$n_{r+1} = n_r^{-1} n_{d+r-1} n_r = n_r^{-1} n_{r-1} n_r.$$

$\square$

We can now prove (7.3), namely that $n_i n_j \ldots = n_j n_i \ldots (d = m_{ij}$ factors). Here $n_i = n_1$, and $n_j = n_d$.

(A.5) PROPOSITION. $n_1 n_d \ldots = n_d n_1 \ldots$ *where each side has* m *factors alternating between* n_1 *and* n_d.

PROOF: By (A.4)(i) $n_d = n_o$. Therefore the left hand side equals $n_1 n_o n_1 \ldots$ $= n_1 n_2 \ldots n_d$ by repeated use of (A.4)(ii). Similarly the right hand side equals $n_o n_1 n_o \ldots = n_1 n_2 \ldots n_d$. $\square$

2. The Natural Labelling for a Moufang Plane.

Let α_1, α_{12}, α_2 be the positive roots, and U_1, U_{12}, U_2 the corresponding root groups in a natural cyclic order in the apartment Σ - see Figure 2.

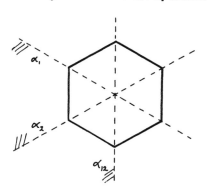

Figure 2

These roots groups are abelian (by (6.12)), and conjugate to one another (e.g., n_1 conjugates U_2 to U_{12}). We shall identify them with a common abelian group A written additively, and use subscripts to indicate membership of U_1, U_{12} or U_2. Moreover A will be given a multiplicative structure, making it an alternative division ring. This will be done via the identification of A with U_{12}, and in such a way that specified non-identity elements $e_1 \in U_1$ and $e_2 \in U_2$ become the unity of A.

Given e_1 and e_2 we write $n_1 = m(e_1)$, $n_2 = m(e_2)$, and set

$$e_{12} = [e_1, e_2].$$

(A.6) LEMMA. $e_{12} = n_2 e_1^{-1} n_2^{-1} = n_1 e_2 n_1^{-1}$.

PROOF: We apply (A.3) for $m = 3$, with e_1^{-1} in place of e_1, and hence n_1^{-1} in place of n_1 by (A.2) (the $e_1, \dots, e_6$ of (A.3) become $e_1, e_{12}, e_2, \dots$). Setting $r = 0$ in (A.3) gives $e_1^{-1} = n_2^{-1} e_{12} n_2$, and setting $r = 1$ in (A.3) gives $e_{12} = n_1 e_2 n_1^{-1}$. □

In view of (A.6) we identify U_1 and U_2 with U_{12} by conjugation, as follows:

$$x_{12} = n_2 x_1^{-1} n_2^{-1} = n_1 x_2 n_1^{-1} \tag{*}$$

Addition on A is multiplication in a root group; since root groups are abelian this is well-defined, and commutative. Multiplication in A is defined via identification with U_{12} as:

$$x * y = [x_1, y_2].$$

In particular $e * e = e$. Before describing the natural blueprint we need the following lemma.

(A.7) LEMMA. (i) $[x_1, y_2^{-1}] = [x_1, y_2]^{-1} = [x_1^{-1}, y_2]$.
(ii) $n_2 x_{12} n_2^{-1} = x_1$ and $n_1 x_{12} n_1^{-1} = x_2^{-1}$.
(iii) $n_1^2 x_2 n_1^{-2} = x_2^{-1}$ and $n_2^2 x_1 n_2^{-2} = x_1^{-1}$.

PROOF: (i) This is a straightforward exercise; it suffices to check the image of the chamber e in Figure 2.

(ii) By (i) if we replace all elements of U_1 and U_2 by their inverses (so, by A.2, n_1 becomes n_1^{-1}, and n_2 becomes n_2^{-1}), the elements of U_{12} remain unchanged ($[x_1^{-1}, y_2^{-1}] = [x_1, y_2]$). The result follows from (*).

(iii) Immediate from (ii). □

Now suppose the sequences (a_1, b_2, c_1) and (x_2, y_1, z_2) are equivalent in the natural blueprint. In other words, following Chapter 7,

$$a_1 n_1 b_2 n_2 c_1 n_1 = x_2 n_2 y_1 n_1 z_2 n_2.$$

Using (*) and (A.7) we can write the left hand side as

$$a_1 b_{12} n_1 c_{12}^{-1} n_2 n_1 = a_1 b_{12} c_2 n_1 n_2 n_1$$

and the right hand side as

$$x_2 y_{12}^{-1} n_2 z_1 n_1 n_2 = x_2 y_{12}^{-1} z_1 n_2 n_1 n_2.$$

Using (A.5) and the fact that U_{12} commutes with U_1 and U_2 , we have

$$a_1 b_{12} c_2 = x_2 y_{12}^{-1} z_1 = y_{12}^{-1} x_2 z_1 = y_{12}^{-1} [x_2, z_1] z_1 x_2 = z_1 y_{12}^{-1} [x_2, z_1] x_2.$$

By uniqueness of the factorization $U = U_1 U_{12} U_2$ we have

$$a_1 = z_1, \ c_2 = x_2, \ b_{12} = y_{12}^{-1} [x_2, z_1] = y_{12}^{-1} [z_1, x_2]^{-1}.$$

Therefore as elements of A,

$$a = z, \ c = x, \ \text{and } y + b = -z * x = -a * c.$$

Interchanging the roles of U_1 and U_2 gives a different multiplication $x *' y = [x_2, y_1^{-1}]$ for which $e *' e = e$. By (A.7)(i) $x *' y = [y_1, x_2] = y * x$. In Chapter 8 we write $(xy)_1$ for $x * y$, and $(xy)_2$ for $x *' y$; with this notation the equivalence of sequences of types 121 and 212 in the natural blueprint may be written:

sequence	type
$x \ y \ z$	1 2 1
$z \ y' \ x$	2 1 2

where $y + y' = (xz)_1 = (zx)_2$.

Exercise. In $SL_3(k)$ identify k with the root groups U_1, U_{12} and U_2 as follows:

$$x_1 = \begin{pmatrix} 1 & x & 0 \\ 0 & 1 & 0 \\ 0 & 0 & 1 \end{pmatrix} \quad x_{12} = \begin{pmatrix} 1 & 0 & x \\ 0 & 1 & 0 \\ 0 & 0 & 1 \end{pmatrix} \quad x_2 = \begin{pmatrix} 1 & 0 & 0 \\ 0 & 1 & x \\ 0 & 0 & 1 \end{pmatrix}.$$

Thus

$$e_1 = \begin{pmatrix} 1 & 1 & 0 \\ 0 & 1 & 0 \\ 0 & 0 & 1 \end{pmatrix}, \ n_1 = \begin{pmatrix} 0 & 1 & 0 \\ -1 & 0 & 0 \\ 0 & 0 & 1 \end{pmatrix}, \text{etc.}$$

Show that

$$a_1 n_1 b_2 n_2 c_1 n_1 = \begin{pmatrix} ac+b & -a & 1 \\ c & -1 & 0 \\ 1 & 0 & 0 \end{pmatrix}$$

and

$$x_2 n_2 y_1 n_1 z_2 n_2 = \begin{pmatrix} -y & -z & 1 \\ x & -1 & 0 \\ 1 & 0 & 0 \end{pmatrix}.$$

Remark. By (6.12) $[x, U_2] = U_{12}$ for $x \in U_1 - \{1\}$, hence every non-zero element of A has a multiplicative inverse. Moreover for $x, y \in U_1$ and $z \in U_2$ one has

$$[x, z].[y, z] = y[x, z]zy^{-1}z^{-1} = yxzx^{-1}y^{-1}z^{-1} = [yx, z],$$

so A satisfies the distributive law, and is therefore a division ring. Moreover it can be shown that A satisfies the alternative laws: $x^2 y = x(xy)$, and $xy^2 = (xy)y$. By the Bruck-Kleinfeld theorem [1951] an alternative division ring is either a field (not necessarily commutative) or a Cayley-Dickson algebra. Thus a Moufang plane is either Desarguesian or is a Cayley plane.

3. The Non-existence Theorem.

The purpose of this section is to prove that there are no Moufang (generalized) d-gons except for $d = 3, 4, 6$ or 8.

This theorem was originally proved by Tits, and the proof appeared in two parts [1976] and [1979]. While part II was in press, a much simpler proof was given by Weiss, using ideas from part I of Tits [1976]. Tits then gave a different, very simple proof using some of Weiss's ideas, and this is what appeared in part II, simultaneously with the paper of Weiss [1979].

The proof given here is based on part II of Tits' paper, with extracts from part I. The main idea is to show first that $1 \neq Z(U) \subset U_i$ where $i = \frac{d+1}{2}$ for d odd, and $i = \frac{d}{2} - 1$ or $\frac{d}{2}$ for d even. One then uses elements $u \in Z(U)$ to obtain inequalities showing that: if d is odd, then $d \leq 3$; if d is 2(mod 4), then $d \leq 6$; and if d is 0(mod 4), then $d \leq 12$. The case $d = 12$ requires further work before a contradiction is reached.

Before going further, we recall that $U = U_1 \ldots U_d$ with uniqueness of decomposition. In particular if $1 \leq i$, $j \leq d$, then $U_{[1,j]} \cap U_{[i,d]} = U_{[i,j]}$ if $i \leq j$, and 1 otherwise. All indices are written mod $2d$, and we shall frequently have occasion to shift our indices, so that, for example, a general relationship between U_j and U_k can be proved by considering U_{i+j} and

U_{i+k}, or U_{-j} and U_{-k}. Notice that if $u \in U_k$, then $m(u)$, acting by conjugation, switches U_j with U_{2k+d-j}. In particular if d is odd, all root groups $U_1, \ldots, U_{2d}$ are conjugate, and if d is even, there are two conjugacy classes: those with even indices, and those with odd indices. To avoid cumbersome notation we shall set $d' = \frac{d}{2}$ for d even, and $\frac{d-1}{2}$ for d odd.

(A.8) LEMMA. *If for some* $1 \le k \le d$, $u \in U_{[k,d]}$, $y \in U_{[1,d]}$ *and* $^y u \in U_{[1,k-1]}$, *then* $u = 1$.

PROOF: Set $y = x^{-1}z$ where $x \in U_{[1,k-1]}$ and $z \in U_{[k,d]}$. Then $^z u \in U_{[k,d]}$, but on the other hand $^z u = {}^x({}^y u) \in U_{[1,k-1]}$, proving that $^z u = 1$. □

(A.9) LEMMA. *Let* $u \in U_i$, $v \in U_j$ *where* $d' + i < j < d + i$, *and let* $x \in U_{[i,j-1]}$. *If* $[vx, u] = 1$, *then* $u = 1$ *or* $v = 1$.

PROOF: Without loss of generality we take $j = d$, to simplify notation. Suppose $v \ne 1$, and let $m = m(v) = wvw'$, where $w, w' \in U_o$.

$$\text{Set } y = {}^m(w'^{-1}x) \in {}^m U_{[o,d-1]} = U_{[1,d]}.$$
$$\text{Since } ym = mw'^{-1}x = wvx, \text{ we have}$$
$$^y({}^m u) = {}^{wvx} u = {}^w u = [w,u]u \in U_{[1,i]}.$$

Moreover $^m u \in U_{d-i} \subset U_{[i+1,d]}$, because $d'+i < j = d$ implies $i \le \frac{d-1}{2}$, and hence $i + 1 \le d - i$. Therefore by the previous lemma $^m u = 1$, and so $u = 1$. □

(A.10) COROLLARY. *Let* $i < j < i + d$, *and suppose* $u \in U_i$ *commutes with* $y \in U_{[i,j]}$. *Then* $j \le i + d'$. □

(A.11) COROLLARY. *Let* $u \in U_1 - \{1\}$, $v \in U_d - \{1\}$. *Then, using C to mean centralizer,*

$$C_U\{u, v\} = \begin{cases} U_{d'+1} & \text{if } d \text{ is odd} \\ U_{d'}U_{d'+1} & \text{if } d \text{ is even.} \end{cases}$$

PROOF: By the previous Corollary, $C_U(u) \subset U_{[1,d'+1]}$, and $C_U(v) \subset U_{[d-d',d]}$. For d odd, $d - d' = d' + 1$, and for d even $d - d' = d'$, so the result follows. □

(A.12) LEMMA. $Z(U) \ne 1$.

PROOF: We first show $Z(U_{[a,b]}) \ne 1$ for some interval $[a, b]$. If U is abelian, there is nothing to prove, so assume non-abelian and let $j < k$ with $k - j$

minimal subject to $[U_j, U_k] \neq 1$ (so $k \geq j+2$). By this minimality assumption U_j and U_k centralize $U_{[j+1,k-1]}$, hence $1 \neq [U_j, U_k] \subset Z(U_{[j+1,k-1]})$ as required.

Now by induction it suffices to show that if $1 \leq s < t < d$ with $Z(U_{[s,t]}) \neq 1$, then $Z(U_{[s,t+1]}) \neq 1$. To prove this write $x \in U$ as $x_1 \ldots x_d$ where $x_i \in U_i$, and define $\lambda(x) = $ least i such that $x_i \neq 1$, and set $\lambda(1) = \infty$. Now let $X = \{x \in Z(U_{[s,t]}) \mid x \neq 1, \lambda(x) \text{ maximal}\}$; we shall show that $X \subset Z(U_{[s,t+1]})$. To prove this, notice first that U_{t+1} normalizes $U_{[s,t]}$, hence normalizes $Z(U_{[s,t]})$. Thus for $x \in X$, $[x, U_{t+1}] \subset Z(U_{[s,t]})$. However for $u \in U_{t+1}$, $\lambda([x, u]) > \lambda(x)$, and therefore $[x, u] = 1$, proving that $x \in Z(U_{[s,t+1]})$, as required. $\quad\square$

(A.13) THEOREM. *For d odd, $1 \neq Z(U) \subset U_{d'+1}$, where $d' = \frac{d-1}{2}$. For d even, $1 \neq Z(U) \subset U_{d'}$ or $U_{d'+1}$, where $d' = \frac{d}{2}$.*

PROOF: By the previous lemma $Z(U) \neq 1$, and for d odd the result is immediate from (A.11). To deal with the case of d even, suppose the result is false. Then by (A.11) we can find

$$x = uv \in U_o U_1 \cap Z(U_{[1-d',d']}) \text{ where } 1 \neq u \in U_o, \ 1 \neq v \in U_1$$

and

$$x' \in U_{d'} U_{d'+1} \cap Z(U), \ x' \notin U_{d'}.$$

Set

$$y = [x, x'] = [u, x'] \in U_{[1,d']}.$$

The fact that the group $U_{[1,d-1]}$ centralizes x', and is normalized by U_o, implies, by an elementary argument, that it centralizes $y = [u, x']$, and therefore also centralizes $[U_o, y]$ and $[U_o, [U_o, y]] \subset U_{[1,d'-2]}$. However by (A.10) the only subgroup of $U_{[1,d'-2]}$ centralized by U_{d-1} is the identity. Therefore $[U_o, [U_o, y]] = 1$.

Thus both U_o and $U_{[1,d-1]}$ centralize $[U_o, y]$, and hence $U_{[o,d-1]}$ centralizes $[U_o, y] \subset U_{[1,d'-1]}$. However $Z(U_{[o,d-1]}) \subset U_{d'-1} U_{d'}$, and we have assumed by way of contradiction that $Z(U_{[o,d-1]}) \not\subset U_{d'-1}$. Therefore $[U_o, y] = 1$.

This, together with the fact (above) that $U_{[1,d-1]}$ centralizes y, shows that

$$y \in Z(U_{[o,d-1]}) \subset U_{d'-1} U_{d'}.$$

Interchanging the roles of x and x' in the argument above gives

$$y^{-1} = [x', x] \in U_1 U_2.$$

Consequently $d' = 2$, and $d = 4$. In this case y is central in $U_{[o,3]}$, and $y = y_1 y_2$ with $y_i \in U_i$. Since y and y_2 centralize U_1 and U_3, so does y_1; but $y_1 \in U_1$ centralizes U_o and U_2, so $y_1 \in Z(U_{[o,3]})$. This contradicts our original assumption, and completes the proof. $\qquad\qquad\qquad\qquad\qquad\square$

The following lemma will be crucial in obtaining bounds on d.

(A.14) LEMMA. *Let $u \in U_i - \{1\}$, and suppose*

$$[u, U_{i-p}] = 1 = [u, U_{i+p}]$$

where $0 < p < \frac{d}{2}$, and p is even, or p and d are both odd. Then $3p \leq d$.

PROOF: Without loss of generality take $i = 0$. Write $m = m(u) = vuv' = u'vu$ where $v, v' \in U_d$ and $u' \in U_o$. Let $x \in U_p$, so $^vx = {}^{vu}x = (^mx)^{u'}$. Now $^mx \in U_{d-p}$, so $^vx = (^mx)^{u'} = [u', {}^mx](^mx)^{-1} \in U_{[1,d-p]}$. Therefore $[x, v] \in U_{[1,d-p]}$, but on the other hand $[U_p, v] \subset U_{[p+1,d-1]}$, so we have

$$[U_p, v] \subset U_{[p+1,d-p]} \qquad\qquad\qquad (1)$$

Similarly:

$$[U_{-p}, v] \subset U_{[-d+p,-p-1]} \qquad\qquad\qquad (2)$$

Now let M_k be the set of elements of N inducing the reflection $\alpha_j \leftrightarrow \alpha_{2k+d-j}$ on our given apartment (e.g. for $z \in U_k$, $m(z) \in M_k$). For p even, let $g \in M_{p/2}$, so $^gU_j = U_{p+d-j}$. For p and d odd, let $g \in M_{(d+p)/2}M_o$, so $^gU_j = U_{p+d+j}$. In both cases $^gv \in U_p$. If p is even apply g to (1), and if p is odd apply g to (2) to obtain:

$$[U_d, {}^gv] \subset U_{[2p,d-1]} \qquad\qquad\qquad (3)$$

Combining (1) and (3) gives:

$$[^gv, v] \in U_{[p+1,d-p]} \cap U_{[2p,d-1]}.$$

Moreover $[^gv, v] \neq 1$ by (A.10) because $^gv \in U_p$, $v \in U_d$, and $d - p > d/2$. Therefore $2p \leq d - p$. $\qquad\qquad\qquad\qquad\qquad\qquad\qquad\qquad\square$

Remark. The case where p is even did not use (2), and therefore only the condition $[u, U_{i+p}] = 1$ was needed.

We are now in a position to prove the main theorem of this section.

(A.15) THEOREM. *For a Moufang d-gon, $d = 3, 4, 6$ or 8.*

PROOF: By (A.13) there exists $u \in Z(U) \subset U_i$ with $u \neq 1$.

(i) If d is odd, then $i = \frac{d+1}{2}$, and $[u, U_{i \pm p}] = 1$ for $p = \frac{d-1}{2}$. Therefore by (A.14) $3p \leq d$. Thus $3d - 3 \leq 2d$, so $d \leq 3$ (and $d = 3$ in this case).

(ii) If d is even, then $i = \frac{d}{2}$ or $\frac{d}{2} + 1$, so $[u, U_{i \pm p}] = 1$ for $p \leq \frac{d}{2} - 1$.
If $d = 2 \pmod 4$, then $p = \frac{d}{2} - 1$ is even, and hence by (A.14) $3\left(\frac{d}{2} - 1\right) \leq d$. Thus $3(d - 2) \leq 2d$, so $d \leq 6$ (and $d = 6$ in this case).
If $d = 0 \pmod 4$, then $p = \frac{d}{2} - 2$ is even, and hence by (A.14) $3\left(\frac{d}{2} - 2\right) \leq d$. Thus $3(d - 4) \leq 2d$, so $d \leq 12$ (and $d = 4, 8$ or 12 in this case).
It remains to deal with the $d = 12$ case.

(iii) $d = 12$ is impossible.

Without loss of generality we may assume $Z(U) \subset U_6$. Since for each $2k$, U_{2k} is conjugate to U_6 by some element of N, the set

$$U_{2k}^{\dagger} = Z(U_{[2k-5, 2k+6]}) - \{1\} \subset U_{2k}$$

is non-empty. With the notation above, $u \in U_6$. We set $m = m(u) = vuv' = u'vu$, where $v, v' \in U_{18}$ and $u' \in U_6$. Given $w \in U_{11} - \{1\}$, it suffices to show that v and w commute, because this contradicts (A.10). The proof that $[w, v] = 1$ will be achieved in two steps, but at one point in the second step we shall need to know that $v \in U_{18}^{\dagger}$; this fact will be proved in Step 3.

Notice first that $[w, v] \in U_{[12, 17]}$.

Step 1. $[w, v] \in U_{[12, 13]}$.

Since u commutes with w, we have

$$^{v}w = \,^{vu}w = (^{m}w)^{u'}.$$

Therefore

$$^{v}w = [u, \,^{m}w]. \,^{m}w \in U_{[7, 13]} \text{ since } ^{m}w \in U_{13}.$$

Hence

$$[w, v] \in U_{[7, 13]}, \text{ and so } [w, v] \in U_{[12, 13]}.$$

Step 2. $[w, v] \in U_{[16, 17]}$.

Let $n = twt' \in M_{11}$, where $t, t' \in U_{23}$. By Step 3, $v \in U_{18}^{\dagger}$ and hence v commutes with t', so we have

$$^{w}v = \,^{wt'}v = (^{n}v)^{t}.$$

Therefore

$$^w v = [t, {}^n v]. \quad {}^n v \in U_{[16,22]} \text{ since } {}^n v \in U_{16}.$$

Hence

$$[w, v] \in U_{[16,17]}.$$

Steps 1 and 2 show $[w, v] = 1$ contradicting (A.10) as required. It remains to prove:

Step 3. $v \in U_{18}^\dagger$.

To prove this take $x \in U_{14}^\dagger$; it suffices to show that v is conjugate to x by an element of N.

Step 3A. $[u, x] = {}^m x$. $\hspace{5cm}$ (A)

Notice first that ${}^m x \in U_{10}$. Since x commutes with $v' \in U_{18}$, we have

$$({}^m x)^v = {}^{uv'} x = {}^u x.$$

Therefore $[u, x] = ({}^m x)^v x^{-1} = {}^m x [{}^m x^{-1}, v] x^{-1} \in {}^m x . U_{[11,17]}$. Moreover $[u, x] \in U_{[7,13]}$, hence $[u, x] \in {}^m x . U_{[11,13]} \subset U_{[10,13]}$. And interchanging the roles of x and u shows $[u, x] \in U_{[7,10]}$. Thus $[u, x] \in U_{[7,10]} \cap {}^m x . U_{[11,13]}$, so $[u, x] = {}^m x$.

Step 3B. Let $y = {}^m x \in U_{10}$. Then $[y^{-1}, v] = x^{-1}$. $\hspace{2.5cm}$ (B)

Indeed using Step 3A for the fourth equality,

$$y = {}^{vuv'} x = {}^{vu} x = {}^v([u, x] x) = {}^v(yx) = {}^v y . x = y[y^{-1}, v] x; \text{ hence}$$
$$[y^{-1}, v] x = 1.$$

Step 3C. $[y^{-1}, {}^{m(y)} x^{-1}] = x^{-1}$. $\hspace{4cm}$ (C)

By (A.2), $m^{-1} = m(u^{-1})$, so $x = {}^{m(u^{-1})} y$, and formula (A) can be rewritten as:

$$[u, {}^{m(u^{-1})} y] = y \text{ where } u \in U_6^\dagger, \ y \in U_{10}^\dagger.$$

In this formula replace $y \in U_{10}$ by $x^{-1} \in U_{14}^\dagger$, and $u \in U_6^\dagger$ by $y^{-1} \in U$ to obtain

$$[y^{-1}, {}^{m(y)} x^{-1}] = x^{-1}, \text{ as required.}$$

Combining (B) and (C) shows $[y^{-1}, {}^{m(y)} x . v] = 1$, and since ${}^{m(y)} x, v \in U_{18}$, and the only element of U_{18} commuting with $y^{-1} \in U_{10}$ is the identity, we have $v = {}^{m(y)} x^{-1}$.

Thus $v \in U^{\dagger}_{18}$ since it is conjugate, via $m(y) \in N$, to an element of $U^{\dagger}_{14}$. This concludes the proof.   □

According to the preceding theorem, Moufang d-gons exist only if $d = 3$, 4, 6, or 8, and it turns out that there is a complete classification in each case. For $d = 3$ the classification is well-known—a Moufang plane is coordinatised by a (skew) field or a Cayley division algebra—and was explained in Section 2 of this appendix. For $d = 4$ the classification is given by Tits and Weiss [2002]; it involves an unexpected family of Moufang quadrangles discovered by Weiss, and he developed a theory of quadrangular algebras to describe this and the other families of exceptional quadrangles—see Weiss [2006]. An alternative description of Weiss's quadrangles by Mühlherr and Van Maldeghem [1999] exhibits them inside F_4 buildings. For $n = 6$ an explicit classification was outlined by Tits [1976a], and a detailed proof, using Jordan algebras, appears in Tits and Weiss [2002]. For $d = 8$ all examples arise from groups of type $^2F_4$, and a complete proof of this fact and an analysis of these groups was given by Tits [1983]. Earlier partial results for the cases $n = 4$ and $n = 6$ were given by Faulkner [1977], and a complete proof in the finite case was given by Fong and Seitz [1973] and [1974].

APPENDIX 2
Diagrams for Moufang Polygons

Moufang Planes.

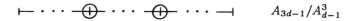

$$A_{3d-1}/A_{d-1}^3$$

As explained in Chapter 8 section 5, this is the diagram for a Desarguesian plane over a field K of finite degree d (dimension d^2) over its centre k; if $d = 1$ this is ⊕——⊕ □

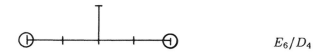

$$E_6/D_4$$

This is the diagram for a Cayley plane, over a division Cayley algebra K. The anisotropic part of the diagram (that obtained by deleting the circled nodes) represents an anisotropic quadratic form (no singular subspaces), namely the reduced norm of K.

Moufang Quadrangles - the Classical Cases.

The "classical" Moufang quadrangles all arise from a (σ, ϵ)-hermitian or pseudo-quadratic form of Witt index 2 on some vector space. For these diagrams it is assumed the vector space has finite dimension N over a field K, which in turn has finite degree d (dimension d^2) over its center k. In all cases except ^2A_n, σ fixes k and so σ is the identity when K is commutative.

$$B_n$$

In this case $K = k$, $\epsilon = 1$, and we have a quadratic form (of Witt index 2) in dimension $N = 2n + 1$. If k is a finite field $n = 2$; if k is p-adic $n = 2$ or 3; if $k = \mathbf{R}$ there is a unique example for each n; and for number fields there is no restriction of n.

If char. $k = 2$ and K is a commutative field such that $K \supset k \supset K^2$ (so k is not perfect) then the fundamental root groups (U_1, U_2) can be associated to (k, K) to provide an exotic form of "mixed type" - see Tits [1976a] (2.5). We assign this the diagram

$$\underset{k \qquad\quad K}{\boxminus\!=\!=\!=\!\boxminus} \qquad\qquad B_2 \text{ mixed}$$

$\square$

$$\vdots \qquad\qquad \left\{ \begin{array}{l} \vdots \\[2mm] \vdots \end{array} \right. \qquad \begin{array}{l} n \text{ even} \\[4mm] n \text{ odd} \end{array} \qquad {}^2A_n$$

One has $d \mid n + 1$, and if $n + 1 = 4d$ the diagram is  Here $[k : k^\sigma] = 2$, $N = \frac{n+1}{d}$, and the form is $(\sigma, 1)$-hermitian. If k is finite $d = 1$ and $n = 3$ or 4; if k is p-adic $d = 1$ and $n = 3$, 4, or 5; if $k^\sigma = \mathbf{R}$, $d = 1$; and if k is a number field there is no special restriction on d or n. $\square$

$$\vdash\!\!-\cdots\!\oplus\!\cdots\!\oplus\!\cdots\!-\!\!=\!\!\Leftarrow\!\!- \qquad\qquad C_n$$

Here $d = 2^s \mid 2n$, and $N = \frac{2n}{d}$. The form is $(\sigma, -1)$-hermitian, and if $d = 1$, then $n = 2$ and the form is symplectic; in this case the diagram is $\oplus\!=\!\Leftarrow\!\ominus$ which is the dual of the B_2 case. If k is finite $d = 1$; if k is p-adic $d = 1$, or $d = 2$ and $n = 4$ or 5; if $k = \mathbf{R}$ or a number field $d = 1$ or 2. $\square$

$$\vdash\!\!-\cdots\!\oplus\!\cdots\!\oplus\!\cdots\!-\!\!\!\prec\!\!\!< \qquad\qquad D_n$$

For $n = 2d$ this is $\vdash\!\!-\cdots\!\oplus\!\cdots\!-\!\!\!\prec\!\!\odot$ and the case $n = 4$, $d = 2$ is the dual of $n = 4$, $d = 1$ $\oplus\!-\!\ominus\!\prec$.

Here $d = 2^s \mid 2n$, $n \neq 2d + 1$ and $N = \frac{2n}{d}$. The form is pseudo-quadratic. For k finite there is no (rank 2) case; if k is p-adic either $n = 4$ (and $d = 1$ or 2 - see diagrams), or $n = 7$ and $d = 2$; if $k = \mathbf{R}$ then $d = 1$ and n is even, or $n = 4$ and $d = 2$; if k is a number field either n is even and $d = 1$ or 2, or $n = 7$ and $d = 2$.   □

^2D_n

If $n = 2d + 1$ then $d = 2$ or 1 and we have 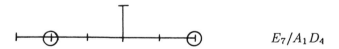 or the latter being the dual of $^2A_3$.

Here again, $d = 2^s \mid 2n$, $N = \frac{2n}{d}$, and the form is pseudo-quadratic; for a given n, the distinction between D_n and ^2D_n depends on the discriminant of the form. If k is finite $d = 1$ and $n = 3$; if k is p-adic either $d = 1$ and $n = 3$, or $d = 2$ and $n = 5$ or 6; if $k = \mathbf{R}$ either $d = 1$ and n is odd, or $d = 2$ and $n = 5$; if k is a number field $d = 1$ or 2.   □

Moufang Quadrangles - the Exceptional Cases.

$^2E_6 / \,^2A_3$

This does not exist over finite fields or p-adic fields, but does exist over the reals and over number fields. The root group dimensions are 6 and 9.   □

E_7/A_1D_4

This does not exist over finite fields, p-adic fields or the reals but does exist over some number fields. The root group dimensions are 17 and 8.   □

E_8/D_6

This does not exist over finite, p-adic or number fields, nor the reals. The root group dimensions are 12 and 33.

$$F_4/B_2$$

These exist within F_4 buildings parameterized by certain non-perfect fields L and L' of characteristic 2 with $L^2 \subset L' \subset L$.

Moufang Hexagons.

$$G_2$$

This arises from a split Cayley algebra, and exists for all fields k; if char. $k = 3$ and $K \supset k \supset K^3$ then the fundamental root groups (U_1, U_2) can be associated to (k, K) to provide an exotic form of "mixed type" - see Tits [1976a] (2.5). We assign it the diagram

$$G_2 \text{ mixed}$$

$$\square$$

$$k \qquad K$$

$$^3D_4 \text{ and } ^6D_4$$

These arise from a building of type D_4, taking the chambers fixed under a triality automorphism involving a field automorphism (if there is no field automorphism one gets G_2). The fundamental root groups (U_1, U_2) can be associated to (k, K) where K is a separable cubic extension of k with Galois group Z_3 or S_3. These cases exist for any field k having the appropriate Galois extension. $\qquad \square$

$$E_6/A_2^2$$

Here (U_1, U_2) can be associated to (k, K) where K is a skew field of degree 3 over its centre k. They exist for all such skew fields (e.g. for k any p-adic field). $\qquad \square$

$$^2E_6/A_2^2$$

Here (U_1, U_2) can be associated to (k, K) where k is a commutative field having a quadratic extension k' over which there is a central division algebra D of degree 3 admitting an involutory automorphism σ such that $k = k'^\sigma$ and $K = D^\sigma$ has dimension 9 over k. They exist for all such situations. $\square$

$$E_8/E_6$$

Here (U_1, U_2) can be associated to (k, J) where J is a 27-dimensional exceptional Jordan division algebra over the commutative field k. They exist for all such Jordan algebras. $\square$

Remark. The existence of these Moufang hexagons is a consequence of an explicit construction given by Tits [1976a].

Moufang Octagons.

$$^2F_4$$

These arise over any commutative field K of characteristic 2 admitting an automorphism σ whose square is the Frobenius (i.e. $\sigma^2 : x \to x^2$). The root groups U_1 and U_2 are isomorphic to K^+, and to the set $K \times K$ with group structure $(t, u).(t', u') = (t + t', u + u' + t't^\sigma)$.

Note. With the exception of $^2F_4$ and the B_2 and G_2 of mixed type, these are the Tits diagrams for simple algebraic groups of relative rank 2 over the field k. They all appear in the general classification given by Tits [1966].

APPENDIX 3
Non-Discrete Buildings

In Chapters 9 and 10 we examined affine buildings and showed how they arise from a group, such as $SL_n(K)$, over a field K having a discrete valuation v. More generally one can consider non-discrete valuations $v :$ $K^\times \to \mathbf{R}$, where $v(ab) = v(a) + v(b)$, and $v(a + b) \geq \min\{v(a), v(b)\}$, in which case Bruhat and Tits [1972] (Chapter 7) define a "non-discrete building" whose "apartments" are affine spaces; it is a topological space, but cannot be regarded as a simplicial complex or chamber system, unless $v(K^\times)$ is discrete. In this brief appendix we shall do little more than give a definition, and discuss the classification of these objects, which will be called *affine apartment systems*. Further details can be obtained from the paper of Tits [1986a], which has already been used extensively in Chapter 10.

First we need some notation. Let $\overline{W}$ be a finite Coxeter group, let V be the vector space of (2.1) on which $\overline{W}$ acts, and let $\mathbf{A}$ be the affine space associated to V. We define W to be the group of affine isometries of $\mathbf{A}$ whose vector part is $\overline{W}$; in other words $\mathbf{R}^n \cdot \overline{W}$ where $\mathbf{R}^n$ is the translation group of $\mathbf{A}$. This notation is exactly that used in [loc. cit.], but notice that W is *not* a Coxeter group; it is different from the W in Chapter 10.

A *wall* of $\mathbf{A}$ means a hyperplane fixed by a reflection of W (in other words a translate of a wall of V); it divides $\mathbf{A}$ into two *half-apartments*. Again this is different from Chapters 9 and 10 because these walls are everywhere dense in $\mathbf{A}$. Similarly one defines *sectors, sector-panels* and *sector-faces* of $\mathbf{A}$ by taking all translates of those in V.

Remark. In Chapters 9 and 10 an affine Coxeter group belonged to an affine diagram, and both $\widetilde{B}_n$ and $\widetilde{C}_n$ give rise to the same $\overline{W}$, of type C_n. The distinction between these cases relies on the distance between adjacent walls in a parallel class. Here however such walls are everywhere dense and there is no affine diagram, only the spherical diagram for $\overline{W}$.

The idea is now to define an object $(\Delta, \mathcal{F})$ which is a set Δ together with a collection $\mathcal{F}$ of injections of $\mathbf{A}$ into Δ satisfying axioms (A1),...,(A5) below. For $f \in \mathcal{F}$, the subset $f(\mathbf{A}) \subset \Delta$ will be called an *apartment* of $(\Delta, \mathcal{F})$, and a *wall, sector*, etc. of $(\Delta, \mathcal{F})$ will mean the image of a wall, sector, etc. of $\mathbf{A}$ under some $f \in \mathcal{F}$. The conditions are:

(A1) If $w \in W$ and $f \in \mathcal{F}$, then $f \circ w \in \mathcal{F}$.

(A2) If $f, f' \in \mathcal{F}$, then $X = f^{-1}(f'(\mathbf{A}))$ is closed and convex in $\mathbf{A}$, and $f|_X = f' \circ w|_X$ for some $w \in W$.

(A3) Any two points of Δ lie in a common apartment.

(A4) Any two sectors contain subsectors lying in a common apartment.

(A5) If A_1, A_2, A_3 are three apartments such that each of $A_1 \cap A_2$, $A_1 \cap A_3$ and $A_2 \cap A_3$ is a half-apartment then $A_1 \cap A_2 \cap A_3 \neq \emptyset$.

Remarks. 1. (A2) and (A3) allow one to define a metric d: given two points p and q of Δ, take $d(p,q)$ to be the Euclidean distance between p and q in any apartment containing both.

2. An alternative to (A5) is:

(A5') Given $f \in \mathcal{F}$ and a point $p \in f(\mathbf{A})$ there is a retraction $\rho : \Delta \to f(\mathbf{A})$ such that $\rho^{-1}(p) = \{p\}$ and the restriction to each apartment diminishes distances.

Both (A5) and (A5') were suggested by Tits as replacements for the (A5) given in Tits [1986a] which, as pointed out by K. Brown, cannot be used in Proposition 17.1 of that paper. In fact the (A5) above was given by Tits in the original lectures on which [loc. cit.] was based; it can be shown to be a consequence of (A5').

Example 1. Take an affine building Δ with a system of apartments $\mathcal{A}$. Treat Δ as a topological space via its simplicial structure, and let $\mathbf{A}$ be the Coxeter complex treated as an affine space. For each $A \in \mathcal{A}$ take an isometry f from $\mathbf{A}$ to A, and let $\mathcal{F}$ denote the set of all $f \circ w$ for $w \in W$ (W being as above, not the Coxeter group). Then $(\Delta, \mathcal{F})$ satisfies (A1) - (A5): in fact (A1) is immediate from the definition of $\mathcal{F}$; (A2) is Exercise 9 of Chapter 9; (A3) and (A4) are immediate from conditions (i) and (ii) for $(\Delta, \mathcal{A})$ at the beginning of Chapter 10; and finally (A5) is Exercise 6 in Chapter 10.

Example 2. $\mathbf{A} = \mathbf{R}$ (i.e. $n = 1$, $\overline{W} \cong Z_2$). Following [loc. cit.] we shall simply call $T = (\Delta, \mathcal{F})$ a *tree* (it is also sometimes called an **R**-tree). Each apartment is a copy of the real line, and two apartments intersect either in the empty set or a closed line segement; in particular between any two points p and q there is a unique line segment of length $d(p,q)$.

As in Chapter 10 section 1, a tree T determines a projective valuation ω_T on its set T^∞ of ends. Conversely the following proposition (combining Propositions 2 and 3 of [loc. cit.]) provides a generalization of 10.2.

(A.16) PROPOSITION. *For any set E having at least 3 elements and a projective valuation ω (in the sense of Chapter 10 section 1), one can identify E with the ends of a tree T such that $\omega = \omega_T$; moreover E and ω determine T up to unique isomorphism.*   □

The proof of uniqueness is given in [loc. cit.] section 16, using a method which can be adapted to prove the existence of T, given ω. The idea is that for each pair $a,b \in E$ one takes a model $A(a,b)$ of the real line, whose points are functions x from $E - \{a,b\}$ to $\mathbf{R}$ satisfying $x(d) - x(c) = w(a,b;c,d)$. The tree is then obtained as the disjoint union of all sets $A(a,b)$, factored out by an equivalence relation.

The Building at Infinity.

As in Chapter 9, one defines two sector-faces to be *parallel* if they are at bounded distance, and it is then straightforward to verify, as in Chapters 9 and 10, that the parallel classes of sector-faces of $(\Delta, \mathcal{F})$ are the simplexes of a spherical building $(\Delta, \mathcal{F})^\infty$ "at infinity". Although $(\Delta, \mathcal{F})$ may be non-discrete, $(\Delta, \mathcal{F})^\infty$ is a building in the usual sense of being a chamber system: its chambers are parallel classes of sectors, and its panels are parallel classes of sector-panels.

Much of the work in Chapter 10 carries through with very little change. As in section 2 of that Chapter, for each wall m of $(\Delta, \mathcal{F})^\infty$, there is a tree $T(m)$ (in the sense of this appendix); its points are the walls M of $(\Delta, \mathcal{F})$ in the direction m, and its ends correspond to the roots of $(\Delta, \mathcal{F})^\infty$ having wall m. Letting $St(m)$ denote this set of roots, $T(m)$ provides a projective valuation ω_m of $St(m)$. Similarly for a panel π of $(\Delta, \mathcal{F})^\infty$ one obtains a projective valuation ω_π on $St(\pi)$. The analogue of (10.5) holds, namely that $(\Delta, \mathcal{F})^\infty$ together with the ω_m or ω_π determines $(\Delta, \mathcal{F})$ up to unique isomorphism.

Also, as in Chapter 10 section 3, if $(\Delta, \mathcal{F})^\infty$ is Moufang, then one obtains a set of root data with valuation (φ_a). Moreover each equivalence class of root data with valuation gives rise to an affine apartment system $(\Delta, \mathcal{F})$, but here the work of Chapter 10 is not sufficient. In the discrete case, section 4 of that Chapter gave an explicit construction of an affine BN-Pair, but in general $(\Delta, \mathcal{F})$ cannot be realized as a chamber system so there is no such BN-Pair. However the construction of $(\Delta, \mathcal{F})$ is given in Chapter 7 of Bruhat-Tits [1972].

If $(\Delta, \mathcal{F})$ has dimension ≥ 3, and the diagram (of $\overline{W}$) is connected, then $(\Delta, \mathcal{F})^{\infty}$ has rank ≥ 3 (and the same diagram) and is therefore Moufang. As in the discrete case we obtain the following theorem.

(A.17) THEOREM. *Every affine apartment system of dimension $n \geq 3$, having a connected diagram, arises from a spherical building of rank n over a field K with a valuation $v : K^{\times} \to \mathbf{R}$. Furthermore these apartment systems are classified by equivalence classes of root data with valuation.* $\square$

In fact root data with valuation can be classified, at least in the case of rank ≥ 3 considered here, and a necessary and sufficient condition can be given for a spherical building over K with valuation v to lead to an affine apartment system. More details are available in Chapter 10, and also of course in Tits [1986a].

APPENDIX 4
Topology and the Steinberg Representation

In Chapter 3 Buildings were defined in terms of chamber systems, and in the finite rank case they can also be regarded as simplicial complexes, and hence acquire a topological structure. In the spherical case each apartment becomes a triangulation of a sphere, and the building has the homotopy type of a bouquet of spheres. In the affine case with a connected diagram each apartment becomes a triangulation of Euclidean space and the building is contractible - see the Theorems below.

However in general the simplicial structure is not necessarily appropriate. For example in the affine case with a non-connected diagram it is better to regard the Coxeter complex as a product of Euclidean spaces (one for each component of the diagram) in which a chamber is a direct product of simplexes. For example the Coxeter complex of type $\widetilde{A}_1 \times \widetilde{A}_1 \times \widetilde{A}_1$ would be the tesselation of $\mathbf{R}^3$ by cubes. A cube has, of course, six faces; these correspond to the six panels of a chamber, opposite faces corresponding to the same $\widetilde{A}_1$ subdiagram. The building in this case would have dimension 3 (though if we treat each chamber as a simplex the dimension is 5); in the terminology of Bruhat-Tits [1972] it is a *polysimplicial complex*. We shall not discuss the general case but refer to Davis [1983], which contains a discussion of topological spaces associated to Coxeter groups, and uses them to construct some interesting aspherical manifolds. For the remainder of this appendix we stick to the spherical and affine case.

Homotopy Type.

If X is a metric space and x is a point of X such that for every point y there is a unique geodesic joining x and y then X is contractible in a very simple way. At time t ($0 \leq t \leq 1$) send y to y_t, where y_t is the point on the unique geodesic from x to y such that $d(x, y_t) = t \cdot d(x, y)$. When this is the case we shall call X *geodesically contractible*.

(A.18) THEOREM. *An affine building Δ is contractible.*

PROOF: Let x be some given point of Δ. If y is any point, the apartments containing x and y intersect in a convex set (Exercise 8 of Chapter 9), and hence there is a unique geodesic from x to y, as this is true in each such apartment. Thus Δ is geodesically contractible.   □

Remark. This theorem and its proof apply equally well to the affine apartment systems of Appendix 3.

(A.19) THEOREM. *A spherical building Δ of rank n is homotopic to a bouquet of $(n-1)$-spheres, and the number of spheres equals the number of chambers opposite a given chamber.*

PROOF: Fix a chamber c and let x be its barycentre. Each apartment A containing c is a sphere; it has a unique chamber c' opposite c, and $A - \{c'\}$ is geodesically contractible to x. Now remove from Δ all chambers opposite c, and call the remaining complex Δ'. Since the intersection of two apartments is convex, Δ' is geodesically contractible to x. Therefore Δ is homotopic to the set of chambers opposite c with their boundaries identified to x. After this identification each chamber becomes a sphere, and the result follows.   □

Homology and the Steinberg Representation.

It follows from Theorem (A.19) that if Δ is a building of spherical type and rank n, then its integral homology is:

$$H_i(\Delta, \mathbf{Z}) = \begin{cases} \mathbf{Z} & \text{if } i = 0 \\ 0 & \text{if } i \neq 0, n-1 \\ \mathbf{Z} \oplus \ldots \oplus \mathbf{Z} & \text{if } i = n-1, \text{ where the number of copies of} \\ & \mathbf{Z} \text{ equals the number of apartments} \\ & \text{containing a given chamber} \end{cases}$$

Now let G be a finite group of Lie type having rank n and characteristic p, and let Δ be its building (see Chapter 8 section 6). Then $H_{n-1}(\Delta)$ provides a representation for G, called the *Steinberg representation*. It was originally discovered in a different form by Steinberg [1956] and [1957], and the interpretation via homology is due to work of Curtis [1966] and Tits and Solomon [1969]. For some applications of this representation, and an extensive list of references, see Humphreys [1987].

To study the action of G on Δ, we regard Δ as a simplicial complex, and let

$$C_{n-1} \xrightarrow{\partial_{n-1}} \ldots \longrightarrow C_1 \xrightarrow{\partial_1} C_0$$

be the associated chain complex. As usual we write $Z_r = \text{Ker } \partial_r$, $B_r = \text{Im } \partial_{r+1}$, $B_n = Z_0 = 0$, and $H_r = Z_r/B_r$. We then let $\gamma_r, \zeta_r, \beta_r$ and η_r be the characters of G on C_r, Z_r, B_r and H_r respectively. Since $H_r = Z_r/B_r$ we have

$$\eta_r = \zeta_r - \beta_r$$

and since $C_r/\text{Ker } \partial_r \cong \text{Im } \partial_r$ we have

$$\gamma_r - \zeta_r = \beta_{r-1}.$$

(A.20) PROPOSITION (HOPF TRACE FORMULA).

$$\sum_{r=0}^{n-1}(-1)^r\gamma_r = \sum_{r=0}^{n-1}(-1)^r\eta_r.$$

PROOF: Indeed $\Sigma(-1)^r(\gamma_r - \eta_r) = \Sigma(-1)^r(\beta_{r-1} + \beta_r) = \beta_{-1} + \beta_n = 0.$ $\square$

If G is a group and H a subgroup, the permutation character of G on cosets of H is denoted 1_H^G. Notice that γ_r is the sum of permutation characters $1_{P_J}^G$ for which P_J corresponds to a face of dimension r (in which case $|J| = n - 1 - r$). Therefore

$$\sum_{r=0}^{n-1}(-1)^r\gamma_r = (-1)^{n-1}\sum_{J \subsetneq I}(-1)^{|J|}1_{P_J}^G.$$

Moreover knowing $H_i(\Delta, \mathbf{Z})$ we have $\eta_0 = 1$, $\eta_i = 0$ for $i \neq 0, n-1$, and $\eta_{n-1} = St$, the Steinberg character. Therefore by (A.20):

$$1 + (-1)^{n-1}St = (-1)^{n-1}\sum_{J \subsetneq I}(-1)^{|J|}1_{P_J}^G.$$

Since $P_I = G$, we have $1_{P_I}^G = 1$, hence

$$St = \sum_{J \subseteq I}(-1)^{|J|}1_{P_J}^G.$$

This formula for the Steinberg character was discovered by Curtis [1966], using Steinberg's original definition of the representation.

(A.21) THEOREM. *The Steinberg representation is irreducible, and if K is a field of characteristic p, then $H_{n-1}(\Delta, K)$ is a projective module for G.*

PROOF: First consider the Coxeter group W acting on the Coxeter complex which we think of as an $(n-1)$-sphere S. Clearly $H_{n-1}(S)$ provides a 1-dimensional (hence irreducible) representation of W; let ϵ denote its character - this is the "reflection character" defined by $\epsilon(r_i) = -1$ for each $i \in I$. As in the case of St above, we have the formula:

$$\epsilon = \sum_{J \subseteq I} (-1)^{|J|} 1^W_{W_J},$$

Furthermore, using (,) for the inner product of characters, we have

$$(1^G_{P_J}, 1^G_{P_K}) = (1^W_{W_J}, 1^W_{W_K}).$$

This is because the inner product of two permutation characters $1^G_{H_1}$ and $1^G_{H_2}$ counts the number of double cosets $H_1 \backslash G / H_2$, and from (5.4) (iv) we have a bijection between $P_J \backslash G / P_K$ and $W_J \backslash W / W_K$. Therefore $(St, St) = (\epsilon, \epsilon)$, and since ϵ is irreducible we have $(\epsilon, \epsilon) = 1$, showing that St is irreducible.

To show that $M = H_{n-1}(\Delta, K)$ is a projective G-module, we first consider it as a U-module M_U. By (6.15) U acts simple-transitively on the set of chambers opposite the chamber c stabilized by U, and hence also on the set of apartments containing c. These apartments form a basis for M, so M_U is a free U-module. Therefore the induced module M_U^G is a free G-module, and it suffices to prove M is a direct summand of M_U^G (this is a standard result in representation theory but we give a direct proof). Let $1 = x_1, \ldots, x_r$ be a set of coset representatives for U in G. The projection $\theta : M_U^G \rightarrow M_U$ sending $\Sigma x_i \otimes m_i$ to m_1 is a U-module homomorphism, and its kernel provides a complement to M_U as a submodule of M_U^G. Since $|G : U| \neq 0$ in K we can define $\tilde{\theta} = \dfrac{1}{|G : U|} \sum_i x_i \theta x_i^{-1}$; it is a G-module homomorphism from M_U^G to M, and its kernel provides a G-module complement for M in M_U^G.   □

Cohomology with Compact Support. By (A.18) an affine building Δ is contractible, so its cohomology $H^i(\Delta)$ is zero for $i > 0$. However Δ is not compact, and the cohomology $H_c^i(\Delta)$ with compact support is not zero. In the locally finite case, Δ is locally compact and it can be compactified by adjoining the building Δ^∞ at infinity, but one must be careful about the topology. Let S be a sector and let $S_1, S_2, \ldots$ be sectors having a sector-panel in common with S, and such that the intersections $S \cap S_1$, $S \cap S_2, \ldots$ become increasingly large, and $\lim_{n \to \infty} S_n = S$.

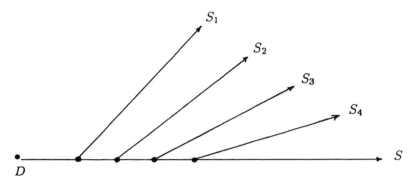

When we compactify Δ by adjoining Δ^∞, we need a topology in which the sequence of chambers $S_1^\infty, S_2^\infty, \ldots$ gets closer and closer to S^∞. Such a topology was given in the $\tilde{A}_1$ case (i.e. when Δ is a tree) in Exercise 10 of Chapter 10, and it can be extended to cases of higher rank. When Δ is locally finite (more precisely if card $St(\pi) \leq$ some finite number s, for all panels π of Δ), this topology makes Δ^∞ compact.

The locally finite case arises from algebraic groups over a local field K, namely $\mathbf{F}_q((t))$ or a p-adic field - see (10.25). This is the case treated by Borel and Serre [1976] who show in their Theorem 5.4 that for $\overline{\Delta} = \Delta \cup \Delta^\infty$ there is a unique topology having the desired properties. The space $\overline{\Delta}$ is compact and contractible; its boundary $\partial\overline{\Delta}$ is Δ^∞ with the topology discussed in the paragraph above.

For cohomology with compact support one has a long exact sequence

$$\ldots \to H_c^i(\overline{\Delta} - \partial\overline{\Delta}) \to \tilde{H}^i(\overline{\Delta}) \to \tilde{H}^i(\partial\overline{\Delta}) \to H_c^{i+1}(\overline{\Delta} - \partial\overline{\Delta}) \to \ldots$$

where $\tilde{H}^i = H^i$ for $i \neq 0$ is "reduced cohomology". The fact that $\overline{\Delta} - \partial\overline{\Delta} = \Delta$, and $\overline{\Delta}$ is contractible (so $\tilde{H}^i(\overline{\Delta}) = 0$) gives:

$$H_c^{i+1}(\Delta) \cong \tilde{H}^i(\partial\overline{\Delta})$$

Borel and Serre [loc. cit.] 2.6 also prove that

$$\tilde{H}^i(\partial\overline{\Delta}) = \begin{cases} C_c^\infty(U; \mathbf{Z}) & \text{if } i = n - 1 \\ 0 & \text{otherwise} \end{cases}$$

where C_c^∞ means C^∞-functions with compact support. Here U can be thought of as a set of points in bijective correspondence with the set of chambers of Δ^∞ opposite a given chamber; it inherits a topology from the topology of $\partial\overline{\Delta}$. Alternatively, think of U as a group, as in Chapter 6 section 4, in which case it acquires a topology as a group of matrices over the locally compact field K. To summarize, we have

$$H_c^i(\Delta) = \begin{cases} C_c^\infty(U;\mathbf{Z}) & \text{if } i = n \\ 0 & \text{otherwise} \end{cases}$$

where Δ is an affine building of dimension n over a local field.

APPENDIX 5
Finite Coxeter Groups (i.e. of spherical type)

	Diagram	$|W|$	Shape of Group (Atlas Notation)
A_n	$\circ\!-\!\!-\!\circ\!-\!\!-\ \cdots\ -\!\!-\!\circ\!-\!\!-\!\circ$	$(n+1)!$	S_{n+1}
C_n	$\circ\!-\!\!-\!\circ\!-\!\!-\ \cdots\ -\!\!-\!\circ\!=\!=\!\circ$	$2^n n!$	$2^n S_n$
D_n		$2^{n-1} n!$	$2^{n-1} S_n$
E_6		$2^7 3^4 5$	$0_6^-(2).2$
E_7		$2^{10} 3^4 5.7$	$2 \times 0_7(2)$
E_8		$2^{14} 3^5 5^2 7$	$2.0_8^+(2).2$
F_4	$\circ\!-\!\!-\!\circ\!=\!=\!\circ\!-\!\!-\!\circ$	1152	$2^3 : S_4 : S_3$
H_3	$\circ\!-\!\!-\!\circ\ \overset{5}{-\!\!-\!\!-}\ \circ$	120	$2 \times A_5$
H_4	$\circ\!-\!\!-\!\circ\!-\!\!-\!\circ\ \overset{5}{-\!\!-\!\!-}\ \circ$	$(120)^2$	$2A_5 : (2 \times A_5)$
$G_2(m)$	$\circ\ \overset{m}{-\!\!-\!\!-\!\!-}\ \circ$	$2m$	D_{2m}

Finite Buildings and Groups of Lie Type

Type of Building	Type of Group	Simple Group (Atlas notation, where different)	Parameters	
A_n	$A_n\,(q)$	$L_{n+1}\,(q)$	$\dfrac{q^n-1}{q-1}$ $\circ \!\!-\!\!\cdots \quad \cdots\!\!-\!\!\circ$ $q \qquad\qquad q$	
C_n	$B_n\,(q)$	$O_{2n+1}\,(q)$	$\dfrac{q^{2n}-1}{q-1}$ $\circ \!\!-\!\!\cdots \quad \cdots\!\!-\!\!\circ \!=\! \circ$ $q \qquad\qquad q \quad q$	$(q+1)(q^2+1)\ldots(q^n+1)$
C_n	$C_n\,(q)$	$S_{2n}\,(q)$	$\dfrac{q^{2n}-1}{q-1}$ $\circ \!\!-\!\!\cdots \quad \cdots\!\!-\!\!\circ \!=\! \circ$ $q \qquad\qquad q \quad q$	$(q+1)(q^2+1)\ldots(q^n+1)$
C_n	$^2A_{2n-1}\,(q)$	$U_{2n}\,(q)$	$\dfrac{(q^{2n-1}+1)(q^{2n}-1)}{q^2-1}$ $\circ \!\!-\!\!\cdots \quad \cdots\!\!-\!\!\circ \!=\! \circ$ $q^2 \qquad\qquad q^2 \quad q$	$(q+1)(q^3+1)\ldots(q^{2n-1}+1)$
C_n	$^2A_{2n}\,(q)$	$U_{2n+1}\,(q)$	$\dfrac{(q^{2n+1}+1)(q^{2n}-1)}{q-1}$ $\circ \!\!-\!\!\cdots \quad \cdots\!\!-\!\!\circ \!=\! \circ$ $q^2 \qquad\qquad q^2 \quad q^3$	$(q^3+1)(q^5+1)\ldots(q^{2n+1}+1)$

Type of Building	Type of Group	Simple Group (Atlas notation, where different)	Parameters		

C_n $\quad {}^2D_{n+1}\,(q)$ $\quad O^-_{2n+2}\,(q)$

$$\frac{(q^{n+1}+1)(q^n-1)}{q-1} \qquad\qquad \frac{(q^{n+1}+1)(q^{2n}-1)}{q^2-1}$$

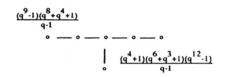

D_n $\quad D_n\,(q)$ $\quad O^+_{2n}\,(q)$

$$\frac{(q^{n-1}+1)(q^n-1)}{q-1} \qquad\qquad (q+1)(q^2+1)\cdots(q^{n-1}+1)$$

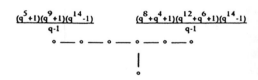

E_6 $\quad E_6\,(q)$

$$\frac{(q^9-1)(q^8+q^4+1)}{q-1}$$

$$\frac{(q^4+1)(q^6+q^3+1)(q^{12}-1)}{q-1}$$

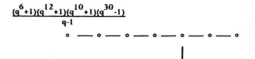

E_7 $\quad E_7\,(q)$

$$\frac{(q^5+1)(q^9+1)(q^{14}-1)}{q-1} \qquad\qquad \frac{(q^8+q^4+1)(q^{12}+q^6+1)(q^{14}-1)}{q-1}$$

E_8 $\quad E_8\,(q)$

$$\frac{(q^6+1)(q^{12}+1)(q^{10}+1)(q^{30}-1)}{q-1}$$

Type of Building	Type of Group	Simple Group (Atlas notation, where different)	Parameters

F_4 $F_4\,(q)$

$$\frac{(q^4+1)(q^{12}-1)}{q-1}$$

$$\underset{q}{\circ} \;\text{---}\; \underset{q}{\circ} \;=\!=\; \underset{q}{\circ} \;\text{---}\; \underset{q}{\circ}$$

F_4 $^2E_6\,(q)$

$$\frac{(q^4+1)(q^6-q^3+1)(q^{12}-1)}{q-1} \qquad\qquad \frac{(q^5+1)(q^9+1)(q^{12}-1)}{q^2-1}$$

$$\underset{q}{\circ} \;\text{---}\; \underset{q}{\circ} \;=\!=\; \underset{q^2}{\circ} \;\text{---}\; \underset{q^2}{\circ}$$

G_2 $G_2\,(q)$

$$(q+1)(q^4+q^2+1)$$

$$\underset{q}{\circ} \;\overset{6}{\text{------------}}\; \underset{q}{\circ}$$

G_2 $^3D_4\,(q)$

$$(q^3+1)(q^8+q^4+1) \qquad\qquad (q+1)(q^8+q^4+1)$$

$$\underset{q^3}{\circ} \;\overset{6}{\text{------------}}\; \underset{q}{\circ}$$

$I_2\,(8)$ $^2F_4\,(q)$

$$(q+1)(q^3+1)(q^6+1) \qquad\qquad (q^2+1)(q^3+1)(q^6+1)$$

$$\underset{q}{\circ} \;\overset{8}{\text{------------}}\; \underset{q^2}{\circ} \qquad q = 2^{\text{odd}}$$

A_1 $^2B_2\,(q)$ $Sz\,(q)$ $\qquad q^2+1$ points $\qquad q = 2^{\text{odd}}$

A_1 $^2G_2\,(q)$ $Ree\,(q)$ $\qquad q^3+1$ points $\qquad q = 3^{\text{odd}}$

The number of chambers per panel is $s+1$, where s is shown below the node of the appropriate type, or if nothing is shown, $s = q$. The number above a node is the number of vertices of the appropriate cotype.

BIBLIOGRAPHY

P. Abramenko, [1996] *Twin Buildings and Applications to S-Arithmetic Groups*, Springer Lect. Notes **1641** (1996).

P. Abramenko and K. S. Brown, [2008] *Buildings - Theory and Applications*, Springer GTM **248**, 2008.

P. Abramenko and H. Van Maldeghem, [1999] Connectedness of opposite-flag geometries in Moufang polygons, *Europ. J. Combinatorics* **20** (1999) 461–468.

P. Abramenko and H. Van Maldeghem, [2001] 1-Twinnings of Buildings, *Math. Zeitschrift* **238** (2001) 187–203.

P. Abramenko and B. Mühlherr, [1997] Présentations de certaines BN-paires jumelées comme sommes amalgamées, *C. R. Acad. Sci. Paris Sér. I Math.* **325** (1997), no. 7, 701–706.

E. Artin, Geometric Algebra, Interscience, New York 1957.

A. Borel and J.-P. Serre, Cohomologie d'immeubles et de groupes S-arithmétiques, *Topology,* **15** (1976), 211–232.

N. Bourbaki, Groupes et Algèbres de Lie, Ch. 4, 5 et 6. Hermann, Paris 1968, Masson, Paris 1981.

K. S. Brown, Buildings, Springer-Verlag 1989.

R. H. Bruck and E. Kleinfeld, The Structure of Alternative Division Rings, *Proc. Amer. Math. Soc.,* **2** (1951), 878–890.

F. Bruhat and J. Tits, Groupes réductifs sur un corps local, I. Données radicielles valuées, *Publ. Math. I.H.E.S.,* **41** (1972). 5–252; II. Schémas en groupes, Existence d'une donnée radicielle valuée, *Publ. Math. I.H.E.S.* **60** (1984), 5–184.

F. Buekenhout and E. E. Shult, On the Foundations of Polar Geometry, *Geometriae Dedicata,* **3** (1974), 155–170.

P.-E. Caprace and B. Mühlherr, [2006] Isomorphisms of Kac-Moody groups which preserve bounded subgroups, *Advances in Math.* **206** (2006) 250–278.

P.-E. Caprace and B. Rémy, [2006] Simplicité abstraite des groupes de Kac-Moody non affines, *C. R. Math. Acad. Sci. Paris* **342** (2006), no. 8, 539–544.

L. Carbone and H. Garland, [2003] Existence of lattices in Kac-Moody groups over finite fields, *Commun. Contemp. Math.* **5** (2003) 813–867.

R. W. Carter, Simple Groups of Lie Type, Wiley-Interscience 1962.

H. S. M. Coxeter, [1934] Discrete Groups generated by Reflections, *Ann. of Math.*, **35** (1934), 588–621.

H. S. M. Coxeter, [1947] Regular Polytopes, 1947, 3rd edition - Dover Publications 1973.

C. W. Curtis, [1964] Groups with a Bruhat decomposition, *Bull Amer. Math. Soc.*, **70** (1964), 357–360.

C. W. Curtis, [1966] The Steinberg character of a finite group with a (B,N)-Pair, *J. Algebra*, **4** (1966), 433–441.

M. W. Davis, Groups generated by reflections and aspherical manifolds not covered by Euclidean space, *Annals of Math.*, **117** (1983), 293–324.

J. R. Faulkner, Steinberg Relations and Coordinization of Polygonal Geometries, *Memoirs Am, Math. Soc.*, vol. 10, no. 185, 1977.

W. Feit and G. Higman, The nonexistence of certain generalized polygons, *J. Algebra*, **1** (1964), 114–131.

W. Feit and J. Tits, Projective representations of minimum degree of group extensions, *Canad. J. Math.*, **30** (1978), 1092–1102.

P. Fong and G. Seitz, Groups with a (B,N)-Pair of Rank 2, I, II, *Inventiones Math.* **21**(1973), 1–57, and **24** (1974), 191–239.

D. Grayson, Finite generation of K-groups of a curve over a finite field, in *Algebraic K-Theory I* (Conference Proc., Oberwolfach 1980), Springer Lecture Notes **966,** (1982), 69–90.

W. Haemers, Eigenvalue Techniques in Design and Graph Theory, Proefschrift, *Mathematisch Centrum,* Amsterdam 1979.

D. Higman, Invariant Relations, Coherent Configurations and Generalized Polygons, pp. 247–263 in *Combinatorics* part 3 (ed. M. Hall and J. Van Lint), Reidel, Dordrecht 1975.

J. E. Humphreys, The Steinberg Representation, *Bull. Amer. Math. Soc.*, **16** (1987), 247–263.

N. Iwahori and H. Matsumoto, On some Bruhat decomposition and the structure of the Hecke ring of a þ-adic Chevalley group, *Publ. I.H.E.S.,* **25** (1965), 5–48.

W. M. Kantor, [1985] Some exceptional 2-adic buildings, *J. Algebra,* **92** (1985), 208–223.

W. M. Kantor, [1986] Generalized Polygons, SCABs and GABs, in *Lecture Notes in Mathematics,* **1181** (Buildings and the Geometry of Diagrams, Como 1984), Springer-Verlag 1986, 79–158.

W. M. Kantor, R. Liebler, and J. Tits, On discrete chamber-transitive automorphism groups of affine buildings, *Bull Amer. Math. Soc.,* **16** (1987), 129–133.

P. Köhler, Th. Meixner and M. Wester, [1984] The affine building of type $\tilde{A}_2$ over a local field of characteristic 2, *Archiv Math.,* **42** (1984), 400–407.

P. Köhler, Th. Meixner and M. Wester, [1985] The 2-adic building of type $\tilde{A}_2$ and its finite projections, *J. Comb. Th. A.,* **38** (1985), 203–209.

V. Landazuri and G. Sietz, On the minimal degrees of projective representations of the finite Chevalley groups, *J. Algebra,* **32** (1974), 418–443.

I.G. Macdonald, Spherical Functions on a Group of þ-adic type, Ramanujan Inst. Publications, **2** (University of Madras), 1971.

R. Moody and K. Teo, Tits systems with crystallographic Weyl groups, *J. Algebra,* **21** (1972), 178–190.

B. Mühlherr, [1998] A Rank 2 Characterization of Twinnings, *Europ. J. Combinatorics* **19** (1998) 603–612.

B. Mühlherr, [1999] Locally split and locally finite twin buildings of 2-spherical type, *J. reine angew. Math.* **511** (1999) 119–143.

B. Mühlherr, [2002] Twin Buildings, in Tits Buildings and the Model Theory of Groups (ed. K. Tent), *London Maths Soc. Lecture Note Series* **291** (2002) 103–117.

B. Mühlherr and M. A. Ronan, [1995] Local to Global Structure in Twin Buildings, *Inventiones math.* **122** (1995) 71–81.

B. Mühlherr and H. Van Maldeghem, [1999] Exceptional Moufang quadrangles of type F_4, *Canad. J. Math.* **51** (1999) 347–371.

B. Rémy, Construction de réseaux en théorie de Kac-Moody, [1999] *C. R. Acad. Sci. Paris Sér. I Math.* **329** (1999), no. 6, 475–478.

B. Rémy and M. Ronan, [2006] Topological groups of Kac-Moody type, Fuchsian twinnings and their lattices, *Comment. Math. Helv.* **81** (2006) 191–219.

M. A. Ronan, [1986] A Construction of Buildings with no Rank 3 Residues of Spherical Type, in *Lecture Notes in Mathematics,* **1181** (Buildings and the Geometry of Diagrams, Como 1984), Springer-Verlag, 1986, 159–190.

M. A. Ronan, [1989] Buildings: Main Ideas and Applications, to appear in *Bull. London Maths. Soc.*

M. A. Ronan, [2000] Local isometries of twin buildings, *Math. Z.* **234** (2000) 435–455.

M. A. Ronan, [2003] Affine Twin Buildings, *J. London Math. Soc.* (*2*) **68** (2003) 461–476.

M. A. Ronan and J. Tits, Building Buildings, *Math. Annalen,* **278** (1987), 291–306.

M. A. Ronan and J. Tits, [1994] Twin Trees I, *Inventiones math.* **116** (1994) 463–479.

M. A. Ronan and J. Tits, [1999] Twin Trees II, *Israel J. Math.* **109** (1999) 349–377.

J.-P. Serre, [1971] Cohomologie des groupes discrets, Prospects in Mathematics, *Ann. of Math. Studies,* **70,** Princeton Univ. Press, Princeton 1971.

J.-P. Serre, [1962/79] Corps Locaux, Hermann, Paris 1962. English translation: Local Fields, Springer-Verlag, 1979.

J.-P. Serre, [1977/80] Arbres, Amalgames, SL_2: *Astérisque,* **46** (1977). English translation: Trees, Springer-Verlag, 1980.

L. Solomon, The Steinberg character of a finite group with a *BN*-Pair, in Theory of Groups (ed. Brauer and Sah), Benjamin, New York 1969, 213–221.

R. Steinberg, Prime power representations of finite linear groups I; II, *Can. J. Math.,* **8** (1956), 580–581; **9** (1957), 347–351.

J. Tits, [1962] Théorème de Bruhat et sous-groupes paraboliques, *C.R. Acad. Sci., Paris Ser. A,* **254** (1962), 2910–2912.

J. Tits, [1966] Classification of algebraic semi-simple groups, in *Proc. Symp. Pure Math.* vol. 9 (Algebraic Groups and Discontinuous Subgroups, Boulder 1965) *Am. Math. Soc.* 1966, 33–62.

J. Tits, [1968] Le Problème des mots dans les groupes de Coxeter. *1st Naz. Alta Mat., Symposia Math.,* **1** (1968). 175–185.

J. Tits, [1974] Buildings of Spherical Type and Finite BN-Pairs, *Lecture Notes in Mathematics,* **386,** Springer-Verlag (1974).

J. Tits, [1976a] Classification of Buildings of Spherical Type and Moufang Polygons: A survey, in *Atti. Coll. Intern. Teorie Combinatorie,* Accad. Naz. Lincei, Rome, 1973 vol. 1 (1976), 229–246.

J. Tits, [1976b] Quadrangles de Moufang, I. Preprint, Paris, 1976.

J. Tits, [1976/79] Non-existence de certains polygones généralisés, I, II. *Inventiones Math.,* **36** (1976), 275–284; **51** (1979), 267–269.

J. Tits, [1977] Endliche Spiegelungsgruppen, die als Weylgruppen auftreten, *Inventiones Math.,* **45** (1977), 283–295.

J. Tits, [1979] Reductive Groups over Local Fields, in *Proc. Symp. Pure Math.* vol. 33 part 1 (Automorphic Forms, Representations and L-Functions, Corvallis 1977) *Am. Math. Soc.,* 1979, 29–69.

J. Tits, [1981] A Local Approach to Buildings, in The Geometric Vein (Coxeter Festschrift), Springer-Verlag, 1981, 317–322.

J. Tits, [1983] Moufang Octagons and the Ree Groups of Type $^2F_4$, *Amer. J. of Math.,* **105** (1983), 539–594.

J. Tits, [1985] Groups and Group Functors attached to Kac-Moody data, *Lecture Notes in Mathematics,* **1111** (Arbeitstagung, Bonn 1984), Springer-Verlag, 1985, 193–223.

J. Tits, [1986a] Immeubles de type affine, in *Lecture Notes in Mathematics,* **1181,** (Buildings and the Geometry of Diagrams, Como 1984), Springer-Verlag, 1986, 159–190.

J. Tits, [1986b] Buildings and Group Amalgamations, *London Maths. Soc. Lecture Notes,* **121** (Proceedings of Groups - St. Andrews 1985), Cambridge Univ. Press, 1986, 110–127.

J. Tits, [1987] Uniqueness and Presentation of Kac-Moody Groups over Fields, *J. Algebra,* **105** (1987), 542–573.

J. Tits, [1992] Twin Buildings and Groups of Kac-Moody Type, in: Liebeck and Saxl (eds.), Groups, Combinatorics and Geometry (Durham 1990), *London Math. Soc. Lecture Note Series* **165**, Cambridge Univ. Press 1992, 249–286.

J. Tits and R. M. Weiss, [2002] *Moufang Polygons*, Springer Monographs in Mathematics 2002.

H. Van Maldeghem, [1987] Non-classical Triangle Buildings, *Geo. Ded.,* **24** (1987), 123–206.

H. Van Maldeghem, [1988] Valuations on PTR's induced by Triangle Buildings, *Geo. Ded.,* **26** (1988), 29–84.

R. Weiss, The Nonexistence of certain Moufang Polygons, *Inventiones Math.,* **51** (1979), 261–266.

R. M. Weiss, [2003] *The Structure of Spherical Buildings*, Princeton Univ. Press 2003.

R. M. Weiss, [2006] *Quadrangular Algebras*, Princeton Univ. Press, Mathematical Notes **46** (2006).

R. M. Weiss, [2008] *The Structure of Affine Buildings*, Princeton Univ. Press, Annals of Mathematics Studies **168** (2009).

E. Witt, Spiegelungsgruppen und Aufzählung halbeinfacher Liescher Ringe, *Hamburger Abhandlungen,* **14** (1941), 289–322.

INDEX OF NOTATION

INDEX